Student's Solutions Manual

for use with

Beginning Algebra

Fifth Edition

James Streeter
Late Professor of Mathematics
Clackamas Community College

Donald Hutchison
Clackamas Community College

Barry Bergman
Clackamas Community College

Louis Hoelzle
Bucks County Community College

Prepared by
Laurel Technical Services

Boston   Burr Ridge, IL   Dubuque, IA   Madison, WI   New York   San Francisco   St. Louis
Bangkok   Bogotá   Caracas   Lisbon   London   Madrid
Mexico City   Milan   New Delhi   Seoul   Singapore   Sydney   Taipei   Toronto

McGraw-Hill Higher Education &

*A Division of The **McGraw-Hill** Companies*

Student's Solutions Manual for use with
BEGINNING ALGEBRA.
JAMES STREETER, DONALD HUTCHISON, BARRY BERGMAN, AND LOUIS HOELZLE.

Published by McGraw-Hill Higher Education, an imprint of The McGraw-Hill Companies, Inc.,
1221 Avenue of the Americas, New York, NY 10020. Copyright © The McGraw-Hill Companies,
Inc., 2001, 1998. All rights reserved.

This book is printed on recycled, acid-free paper containing 10% postconsumer waste.

RECYCLED

7 8 9 0 QPD QPD 0 3

ISBN 007-237718-6

www.mhhe.com

Table of Contents

Chapter 0   An Arithmetic Review

Exercises 0.1

1. $1 \cdot 4 = 4$
$2 \cdot 2 = 4$
Factors of 4: 1, 2, 4

3. $1 \cdot 10 = 10$
$2 \cdot 5 = 10$
Factors of 10: 1, 2, 5, 10

5. $1 \cdot 15 = 15$
$3 \cdot 5 = 15$
Factors of 15: 1, 3, 5, 15

7. $1 \cdot 24 = 24$
$2 \cdot 12 = 24$
$3 \cdot 8 = 24$
$4 \cdot 6 = 24$
Factors of 24: 1, 2, 3, 4, 6, 8, 12, 24

9. $1 \cdot 64 = 64$
$2 \cdot 32 = 64$
$4 \cdot 16 = 64$
$8 \cdot 8 = 64$
Factors of 64: 1, 2, 4, 8, 16, 32, 64

11. $1 \cdot 11 = 11$
Factors of 11: 1, 11

For exercises 13–15, a *sieve of Eratosthenes* may be helpful

	2	3	4	5	6	7	8	9	10
11	12	13	14	15	16	17	18	19	20
21	22	23	24	25	26	27	28	29	30
31	32	33	34	35	36	37	38	39	40
41	42	43	44	45	46	47	48	49	50
51	52	53	54	55	56	57	58	59	60
61	62	63	64	65	66	67	68	69	70
71	72	73	74	75	76	77	78	79	80
81	82	83	84	85	86	87	88	89	90
91	92	93	94	95	96	97	98	99	100
101	102	103	104	105	106	107	108	109	110

13. 19, 23, 31, 59, 97, 103

15. 31, 37, 41, 43, 47

17. $18 = 3 \cdot 6 = 3 \cdot 2 \cdot 3 = 2 \cdot 3 \cdot 3$

19. $30 = 5 \cdot 6 = 5 \cdot 2 \cdot 3 = 2 \cdot 3 \cdot 5$

21. $51 = 3 \cdot 17$

23. $63 = 7 \cdot 9 = 7 \cdot 3 \cdot 3 = 3 \cdot 3 \cdot 7$

25. $70 = 7 \cdot 10 = 7 \cdot 2 \cdot 5 = 2 \cdot 5 \cdot 7$

27. $66 = 6 \cdot 11 = 2 \cdot 3 \cdot 11$

29. $130 = 10 \cdot 13 = 2 \cdot 5 \cdot 13$

31. $315 = 9 \cdot 35 = 3 \cdot 3 \cdot 5 \cdot 7$

33. $225 = 9 \cdot 25 = 3 \cdot 3 \cdot 5 \cdot 5$

35. $189 = 9 \cdot 21 = 3 \cdot 3 \cdot 3 \cdot 7$

37. $1 \cdot 24 = 24$
$2 \cdot 12 = 24$
$3 \cdot 8 = 24$
$\mathbf{4 \cdot 6 = 24}$
$4 + 6 = 10; \; 4, 6$

39. $1 \cdot 30 = 30$
$2 \cdot 15 = 30$
$3 \cdot 10 = 30$
$\mathbf{5 \cdot 6 = 30}$
$\mathbf{6 - 5} = 1; \; 5, 6$

41. Factor of 4: **1, 2**, 4
Factors of 6: **1, 2**, 3, 6
GCF: 2

43. Factors of 10: **1**, 2, **5**, 10
Factors of 15: **1**, 3, **5**, 15
GCF: 5

45. Factors of 21: **1, 3**, 7, 21
Factors of 24: **1**, 2, **3**, 4, 6, 8, 12, 24
GCF: 3

47. Factors of 20: **1**, 2, 4, 5, 10, 20
Factors of 21: **1**, 3, 7, 21
GCF: 1

49. Factors of 18: **1, 2, 3, 6**, 9, 18
Factors of 24: **1, 2, 3**, 4, **6**, 8, 12, 24
GCF: 6

51. Factors of 18: **1, 2, 3, 6, 9, 18**
Factors of 54: **1, 2, 3, 6, 9, 18**, 27, 54
GCF: 18

53. Factors of 36: **1, 2, 3, 4, 6**, 9, **12**, 18, 36
Factors of 48: **1, 2, 3, 4, 6**, 8, **12**, 16, 24, 48
GCF: 12

55. $84 = 4 \cdot 21 = 2 \cdot 2 \cdot \mathbf{3} \cdot \mathbf{7}$
$105 = 15 \cdot 7 = \mathbf{3} \cdot 5 \cdot \mathbf{7}$
GCF $= 3 \cdot 7 = 21$

57. $45 = 9 \cdot 5 = \mathbf{3} \cdot 3 \cdot \mathbf{5}$
$60 = 6 \cdot 10 = 2 \cdot \mathbf{3} \cdot 2 \cdot \mathbf{5} = 2 \cdot 2 \cdot \mathbf{3} \cdot \mathbf{5}$
$75 = 3 \cdot 25 = \mathbf{3} \cdot \mathbf{5} \cdot 5$
GCF $= 3 \cdot 5 = 15$

59. $12 = 4 \cdot 3 = \mathbf{2 \cdot 2 \cdot 3}$
$36 = 4 \cdot 9 = \mathbf{2 \cdot 2 \cdot 3} \cdot 3$
$60 = 6 \cdot 10 = \mathbf{2 \cdot 3} \cdot 2 \cdot 5 = \mathbf{2 \cdot 2 \cdot 3} \cdot 5$
GCF $= 2 \cdot 2 \cdot 3 = 12$

61. $105 = 5 \cdot 21 = 5 \cdot 3 \cdot 7 = 3 \cdot \mathbf{5} \cdot \mathbf{7}$
$140 = 10 \cdot 14 = 2 \cdot \mathbf{5} \cdot 2 \cdot \mathbf{7} = 2 \cdot 2 \cdot \mathbf{5} \cdot \mathbf{7}$
$175 = 7 \cdot 25 = 7 \cdot 5 \cdot 5 = \mathbf{5} \cdot 5 \cdot \mathbf{7}$
GCF $= 5 \cdot 7 = 35$

63. $25 = \mathbf{5 \cdot 5}$
$75 = 3 \cdot 25 = 3 \cdot \mathbf{5 \cdot 5}$
$150 = 10 \cdot 15 = 2 \cdot \mathbf{5} \cdot 3 \cdot \mathbf{5} = 2 \cdot 3 \cdot \mathbf{5 \cdot 5}$
GCF $= 5 \cdot 5 = 25$

65. Challenge exercise

67. Challenge exercise

69. Writing exercise

71. Group exercise

Exercises 0.2

1. $\dfrac{3}{7} \cdot \dfrac{2}{2} = \dfrac{6}{14}; \; \dfrac{3}{7} \cdot \dfrac{3}{3} = \dfrac{9}{21}; \; \dfrac{3}{7} \cdot \dfrac{4}{4} = \dfrac{12}{28};$
$\dfrac{6}{14}, \dfrac{9}{21}, \dfrac{12}{28}$

3. $\dfrac{4}{9} \cdot \dfrac{2}{2} = \dfrac{8}{18}; \; \dfrac{4}{9} \cdot \dfrac{4}{4} = \dfrac{16}{36}; \; \dfrac{4}{9} \cdot \dfrac{10}{10} = \dfrac{40}{90};$
$\dfrac{8}{18}, \dfrac{16}{36}, \dfrac{40}{90}$

5. $\dfrac{5}{6} \cdot \dfrac{2}{2} = \dfrac{10}{12}$; $\dfrac{5}{6} \cdot \dfrac{3}{3} = \dfrac{15}{18}$; $\dfrac{5}{6} \cdot \dfrac{10}{10} = \dfrac{50}{60}$;

$\dfrac{10}{12}, \dfrac{15}{18}, \dfrac{50}{60}$

7. $\dfrac{10}{17} \cdot \dfrac{2}{2} = \dfrac{20}{34}$; $\dfrac{10}{17} \cdot \dfrac{3}{3} = \dfrac{30}{51}$; $\dfrac{10}{17} \cdot \dfrac{10}{10} = \dfrac{100}{170}$;

$\dfrac{20}{34}, \dfrac{30}{51}, \dfrac{100}{170}$

9. $\dfrac{9}{16} \cdot \dfrac{2}{2} = \dfrac{18}{36}$; $\dfrac{9}{16} \cdot \dfrac{3}{3} = \dfrac{27}{48}$; $\dfrac{9}{16} \cdot \dfrac{10}{10} = \dfrac{90}{160}$;

$\dfrac{18}{36}, \dfrac{27}{48}, \dfrac{90}{160}$

11. $\dfrac{7}{9} \cdot \dfrac{2}{2} = \dfrac{14}{18}$; $\dfrac{7}{9} \cdot \dfrac{5}{5} = \dfrac{35}{45}$; $\dfrac{7}{9} \cdot \dfrac{20}{20} = \dfrac{140}{180}$;

$\dfrac{14}{18}, \dfrac{35}{45}, \dfrac{140}{180}$

13. $\dfrac{8}{12} = \dfrac{2 \cdot 2 \cdot 2}{2 \cdot 2 \cdot 3} = \dfrac{2}{3}$

15. $\dfrac{10}{14} = \dfrac{2 \cdot 5}{2 \cdot 7} = \dfrac{5}{7}$

17. $\dfrac{12}{18} = \dfrac{2 \cdot 2 \cdot 3}{2 \cdot 3 \cdot 3} = \dfrac{2}{3}$

19. $\dfrac{35}{40} = \dfrac{5 \cdot 7}{2 \cdot 2 \cdot 2 \cdot 5} = \dfrac{7}{8}$

21. $\dfrac{11}{44} = \dfrac{1 \cdot 11}{2 \cdot 2 \cdot 11} = \dfrac{1}{4}$

23. $\dfrac{12}{36} = \dfrac{2 \cdot 2 \cdot 3}{2 \cdot 2 \cdot 3 \cdot 3} = \dfrac{1}{3}$

25. $\dfrac{24}{27} = \dfrac{2 \cdot 2 \cdot 2 \cdot 3}{3 \cdot 3 \cdot 3} = \dfrac{8}{9}$

27. $\dfrac{32}{40} = \dfrac{2 \cdot 2 \cdot 2 \cdot 2 \cdot 2}{2 \cdot 2 \cdot 2 \cdot 5} = \dfrac{4}{5}$

29. $\dfrac{75}{105} = \dfrac{3 \cdot 5 \cdot 5}{3 \cdot 5 \cdot 7} = \dfrac{5}{7}$

31. $\dfrac{48}{60} = \dfrac{2 \cdot 2 \cdot 2 \cdot 2 \cdot 3}{2 \cdot 2 \cdot 3 \cdot 5} = \dfrac{4}{5}$

33. $\dfrac{105}{135} = \dfrac{3 \cdot 5 \cdot 7}{3 \cdot 3 \cdot 3 \cdot 5} = \dfrac{7}{9}$

35. $\dfrac{15}{44} = \dfrac{3 \cdot 5}{2 \cdot 2 \cdot 11} = \dfrac{15}{44}$

37. $\dfrac{3}{4} \cdot \dfrac{7}{5} = \dfrac{3 \cdot 7}{2 \cdot 2 \cdot 5} = \dfrac{21}{20}$

39. $\dfrac{3}{5} \cdot \dfrac{5}{7} = \dfrac{3 \cdot 5}{5 \cdot 7} = \dfrac{3}{7}$

41. $\dfrac{6}{13} \cdot \dfrac{4}{9} = \dfrac{2 \cdot 3 \cdot 2 \cdot 2}{13 \cdot 3 \cdot 3} = \dfrac{2 \cdot 2 \cdot 2}{13 \cdot 3} = \dfrac{8}{39}$

43. $\dfrac{3}{11} \cdot \dfrac{7}{9} = \dfrac{3 \cdot 7}{11 \cdot 3 \cdot 3} = \dfrac{7}{11 \cdot 3} = \dfrac{7}{33}$

45. $\dfrac{3}{10} \cdot \dfrac{5}{9} = \dfrac{3 \cdot 5}{2 \cdot 5 \cdot 3 \cdot 3} = \dfrac{1}{2 \cdot 3} = \dfrac{1}{6}$

47. $\dfrac{1}{5} \div \dfrac{3}{4} = \dfrac{1}{5} \cdot \dfrac{4}{3} = \dfrac{2 \cdot 2}{3 \cdot 5} = \dfrac{4}{15}$

49. $\dfrac{2}{5} \div \dfrac{3}{4} = \dfrac{2}{5} \cdot \dfrac{4}{3} = \dfrac{2 \cdot 2 \cdot 2}{3 \cdot 5} = \dfrac{8}{15}$

51. $\dfrac{8}{9} \div \dfrac{4}{3} = \dfrac{8}{9} \cdot \dfrac{3}{4} = \dfrac{2 \cdot 2 \cdot 2 \cdot 3}{3 \cdot 3 \cdot 2 \cdot 2} = \dfrac{2}{3}$

53. $\dfrac{7}{10} \div \dfrac{5}{9} = \dfrac{7}{10} \cdot \dfrac{9}{5} = \dfrac{7 \cdot 3 \cdot 3}{2 \cdot 5 \cdot 5} = \dfrac{63}{50}$

55. $\dfrac{8}{15} \div \dfrac{2}{5} = \dfrac{8}{15} \cdot \dfrac{5}{2} = \dfrac{2 \cdot 2 \cdot 2 \cdot 5}{3 \cdot 5 \cdot 2} = \dfrac{2 \cdot 2}{3} = \dfrac{4}{3}$

57. $\dfrac{5}{27} \div \dfrac{25}{36} = \dfrac{5}{27} \cdot \dfrac{36}{25}$

$= \dfrac{5 \cdot 2 \cdot 2 \cdot 3 \cdot 3}{3 \cdot 3 \cdot 3 \cdot 5 \cdot 5}$

$= \dfrac{2 \cdot 2}{3 \cdot 5}$

$= \dfrac{4}{15}$

59. $\dfrac{2}{5} + \dfrac{1}{4} = \dfrac{8}{20} + \dfrac{5}{20} = \dfrac{13}{20}$

61. $\dfrac{2}{5} + \dfrac{7}{15} = \dfrac{6}{15} + \dfrac{7}{15} = \dfrac{13}{15}$

63. $\dfrac{3}{8} + \dfrac{5}{12} = \dfrac{9}{24} + \dfrac{10}{24} = \dfrac{19}{24}$

65. $\dfrac{2}{15}+\dfrac{9}{20}=\dfrac{8}{60}+\dfrac{27}{60}=\dfrac{35}{60}=\dfrac{5\cdot 7}{2\cdot 2\cdot 3\cdot 5}=\dfrac{7}{12}$

67. $\dfrac{7}{15}+\dfrac{13}{18}=\dfrac{42}{90}+\dfrac{65}{90}=\dfrac{107}{90}$

69. $\dfrac{1}{2}+\dfrac{1}{4}+\dfrac{1}{8}=\dfrac{4}{8}+\dfrac{2}{8}+\dfrac{1}{8}=\dfrac{7}{8}$

71. $\dfrac{8}{9}-\dfrac{3}{9}=\dfrac{5}{9}$

73. $\dfrac{5}{8}-\dfrac{1}{8}=\dfrac{4}{8}=\dfrac{1}{2}$

75. $\dfrac{7}{8}-\dfrac{2}{3}=\dfrac{21}{24}-\dfrac{16}{24}=\dfrac{5}{24}$

77. $\dfrac{11}{18}-\dfrac{2}{9}=\dfrac{11}{18}-\dfrac{4}{18}=\dfrac{7}{18}$

79. $\dfrac{5}{8}-\dfrac{1}{6}=\dfrac{15}{24}-\dfrac{4}{24}=\dfrac{11}{24}$

81. $\dfrac{8}{21}-\dfrac{1}{14}=\dfrac{16}{42}-\dfrac{3}{42}=\dfrac{13}{42}$

83. $\begin{array}{r}7.1562\\ +14.78\\ \hline 21.9362\end{array}$

85. $\begin{array}{r}11.12\\ +8.3792\\ \hline 19.4992\end{array}$

87. $\begin{array}{r}9.20\\ -2.85\\ \hline 6.35\end{array}$

89. $\begin{array}{r}18.234\\ -13.64\\ \hline 4.594\end{array}$

91. $\begin{array}{r}3.21\\ \times 2.1\\ \hline 321\\ 6420\\ \hline 6.741\end{array}$

93. $\begin{array}{r}6.29\\ \times 9.13\\ \hline 1887\\ 6290\\ 566100\\ \hline 57.4277\end{array}$

95. $\dfrac{1}{2}+\dfrac{1}{3}+\dfrac{1}{4}=\dfrac{6}{12}+\dfrac{4}{12}+\dfrac{3}{12}=\dfrac{13}{12}$ yd

97. $\dfrac{2}{3}\cdot 240=\dfrac{2\cdot 3\cdot 80}{3}=\$160$

99. $\dfrac{3}{8}\cdot\dfrac{200}{1}=\dfrac{3\cdot 8\cdot 25}{8}=75$ mi

101. $\dfrac{3}{4}\cdot 80=\dfrac{240}{4}=60$ in.

103. $\dfrac{3}{4}\cdot\dfrac{5}{9}=\dfrac{3\cdot 5}{2\cdot 2\cdot 3\cdot 3}=\dfrac{5}{2\cdot 2\cdot 3}=\dfrac{5}{12}$

105. $3\dfrac{1}{3}\cdot 3\dfrac{3}{4}=\dfrac{10}{3}\cdot\dfrac{15}{4}=\dfrac{2\cdot 5\cdot 3\cdot 5}{3\cdot 2\cdot 2}=\dfrac{5\cdot 5}{2}=\dfrac{25}{2}$

107. $540\cdot 4\dfrac{2}{3}=\dfrac{540}{1}\cdot\dfrac{14}{3}=\dfrac{3\cdot 180\cdot 14}{3}$
$=180\cdot 14=2520$ mi

109. $21\cdot\dfrac{22}{7}=\dfrac{3\cdot 7\cdot 2\cdot 11}{7}=3\cdot 2\cdot 11=66$ in.

111. $2\dfrac{1}{4}\cdot 3\dfrac{7}{8}\cdot 4\dfrac{5}{6}=\dfrac{9}{4}\cdot\dfrac{31}{8}\cdot\dfrac{29}{6}$
$=\dfrac{3\cdot 3\cdot 31\cdot 29}{2\cdot 2\cdot 2\cdot 2\cdot 2\cdot 2\cdot 3}$
$=\dfrac{3\cdot 31\cdot 29}{2\cdot 2\cdot 2\cdot 2\cdot 2\cdot 2}$
$=\dfrac{2697}{64}$
$=42\dfrac{9}{64}$ in.^{3}

113. Challenge exercise

Exercises 0.3

1. $7\cdot 7\cdot 7\cdot 7=7^4$

3. $6\cdot 6\cdot 6\cdot 6\cdot 6=6^5$

5. $8\cdot 8\cdot 8\cdot 8\cdot 8\cdot 8\cdot 8\cdot 8\cdot 8\cdot 8=8^{10}$

7. $15 \cdot 15 \cdot 15 \cdot 15 \cdot 15 \cdot 15 = 15^6$

9. $7 + 2 \cdot 6 = 7 + 12 = 19$

11. $(7 + 2) \cdot 6 = 9 \cdot 6 = 54$

13. $12 - 8 \div 4 = 12 - 2 = 10$

15. $(12 - 8) \div 4 = 4 \div 4 = 1$

17. $8 \cdot 7 + 2 \cdot 2 = 56 + 4 = 60$

19. $8 \cdot (7 + 2) \cdot 2 = 8 \cdot 9 \cdot 2 = 144$

21. $3 \cdot 5^2 = 3 \cdot 25 = 75$

23. $(3 \cdot 5)^2 = 15^2 = 225$

25. $4 \cdot 3^2 - 2 = 4 \cdot 9 - 2 = 36 - 2 = 34$

27. $7 \cdot (2^3 - 5) = 7 \cdot (8 - 5) = 7 \cdot 3 = 21$

29. $3 \cdot 2^4 - 6 \cdot 2 = 3 \cdot 16 - 12 = 48 - 12 = 36$

31. $(2 \cdot 4)^2 - 8 \cdot 3 = 8^2 - 8 \cdot 3 = 64 - 24 = 40$

33. $4 \cdot (2 + 6)^2 = 4 \cdot 8^2 = 4 \cdot 64 = 256$

35. $(4 \cdot 2 + 6)^2 = (8 + 6)^2 = 14^2 = 196$

37. $3 \cdot (4 + 3)^2 = 3 \cdot 7^2 = 3 \cdot 49 = 147$

39. $3 \cdot 4 + 3^2 = 12 + 9 = 21$

41.
$$\begin{aligned}
4 \cdot (2 + 3)^2 - 25 &= 4 \cdot 5^2 - 25 \\
&= 4 \cdot 25 - 25 \\
&= 100 - 25 \\
&= 75
\end{aligned}$$

43.
$$\begin{aligned}
(4 \cdot 2 + 3)^2 - 25 &= (8 + 3)^2 - 25 \\
&= 11^2 - 25 \\
&= 121 - 25 \\
&= 96
\end{aligned}$$

45. 1.2

47. 7.8

49. 2^5

51. $36 \div (4 + 2) - 4 = 36 \div 6 - 4 = 6 - 4 = 2$

53.
$$\begin{aligned}
(6 + 9) \div 3 + (16 - 4) \cdot 2 &= 15 \div 3 + 12 \cdot 2 \\
&= 5 + 24 \\
&= 29
\end{aligned}$$

Section 0.4

1. +400

3. −200

5. −25,000

7.

9. 5, 175, −234

11. −7, −5, −1, 0, 2, 3, 8

13. −11, −6, −2, 1, 4, 5, 9

15. −7, −6, −3, 3, 6, 7

17. Maximum: 15
Minimum: −6

19. Maximum: 21
Minimum: −15

21. Maximum: 5
Minimum: −2

23. −15

25. −11

27. 19

29. 7

31. $|17| = 17$

33. $|-10| = 10$

35. $-|3| = -3$

37. $-|-8| = -8$

39. $|-2| + |3| = 2 + 3 = 5$

41. $|-9| + |9| = 9 + 9 = 18$

43. $|4| - |-4| = 4 - 4 = 0$

45. $|15| - |8| = 7$

47. $|15 - 8| = |7| = 7$

49. $|-9| + |2| = 9 + 2 = 11$

51. $|-8| - |-7| = 8 - 7 = 1$

53. True

55. False; –6 is an integer that is not a whole number.

57. False; –4 is a negative integer that is not a whole number.

59. $|6 + (-2)| = |6 - 2| = |4| = 4$

61. $|6| + |-2| = 6 + 2 = 8$

63. –5 cm

65. –$50

67. –10°F

69. –8

71. +$90,000,000

73. a. –6

  b. 8

  c. 8

  d. –2

75. a. –2

  b. 6

  c. 6

  d. 0

77. Writing exercise

Summary Exercises for Chapter 0

1. $1 \cdot 52 = 52$
  $2 \cdot 26 = 52$
  $4 \cdot 13 = 52$
  Factors of 52: 1, 2, 4, 13, 26, 52

3. $1 \cdot 76 = 76$
  $2 \cdot 38 = 76$
  $4 \cdot 19 = 76$
  Factors of 76: 1, 2, 4, 19, 38, 76

5. Prime: 2, 5, 7, 11, 17, 23, 43
  Composite: 14, 21, 27, 39

7. $420 = 20 \cdot 21$
  $= 4 \cdot 5 \cdot 3 \cdot 7$
  $= 2 \cdot 2 \cdot 5 \cdot 3 \cdot 7$
  $= 2^2 \cdot 3 \cdot 5 \cdot 7$

9. $180 = 9 \cdot 20 = 3 \cdot 3 \cdot 4 \cdot 5 = 3 \cdot 3 \cdot 2 \cdot 2 \cdot 5 = 2^2 \cdot 3^2 \cdot 5$

11. Factors of 30: 1, 2, 3, 5, 6, 10, 15, 30
  Factors of 31: 1, 31
  GCF: 1

13. $240 = 12 \cdot 20$
  $= 3 \cdot 4 \cdot 4 \cdot 5$
  $= 3 \cdot 2 \cdot 2 \cdot 2 \cdot 2 \cdot 5$
  $= \mathbf{2} \cdot \mathbf{2} \cdot 2 \cdot 2 \cdot \mathbf{3} \cdot \mathbf{5}$
  $900 = 9 \cdot 100$
  $= 3 \cdot 3 \cdot 10 \cdot 10$
  $= 3 \cdot 3 \cdot 2 \cdot 5 \cdot 2 \cdot 5$
  $= \mathbf{2} \cdot \mathbf{2} \cdot \mathbf{3} \cdot 3 \cdot \mathbf{5} \cdot 5$
  $GCF = 2 \cdot 2 \cdot 3 \cdot 5 = 60$

15. $\dfrac{3}{11} \cdot \dfrac{2}{2} = \dfrac{6}{22}; \ \dfrac{3}{11} \cdot \dfrac{3}{3} = \dfrac{9}{33}; \ \dfrac{3}{11} \cdot \dfrac{4}{4} = \dfrac{12}{44};$
  $\dfrac{6}{22}, \dfrac{9}{33}, \dfrac{12}{44}$

17. $\dfrac{24}{64} = \dfrac{2 \cdot 2 \cdot 2 \cdot 3}{2 \cdot 2 \cdot 2 \cdot 2 \cdot 2 \cdot 2} = \dfrac{3}{2 \cdot 2 \cdot 2} = \dfrac{3}{8}$

19. $\dfrac{10}{27} \cdot \dfrac{9}{20} = \dfrac{2 \cdot 5 \cdot 3 \cdot 3}{3 \cdot 3 \cdot 3 \cdot 2 \cdot 2 \cdot 5} = \dfrac{1}{3 \cdot 2} = \dfrac{1}{6}$

21. $\dfrac{7}{15} \div \dfrac{14}{25} = \dfrac{7}{15} \cdot \dfrac{25}{14} = \dfrac{5 \cdot 5 \cdot 7}{2 \cdot 3 \cdot 5 \cdot 7} = \dfrac{5}{2 \cdot 3} = \dfrac{5}{6}$

23. $\dfrac{5}{18} + \dfrac{7}{12} = \dfrac{10}{36} + \dfrac{21}{36} = \dfrac{31}{36}$

25. $\dfrac{11}{27} - \dfrac{5}{18} = \dfrac{22}{54} - \dfrac{15}{54} = \dfrac{7}{54}$

27. $\begin{array}{r} 10.127 \\ -\,5.49 \\ \hline 4.637 \end{array}$

29. $18 - 3 \cdot 5 = 18 - 15 = 3$

31. $5 \cdot 4^2 = 5 \cdot 16 = 80$

33. $5 \cdot 3^2 - 4 = 5 \cdot 9 - 4 = 45 - 4 = 41$

35. $5 \cdot (4 - 2)^2 = 5 \cdot 2^2 = 5 \cdot 4 = 20$

37. $(5 \cdot 4 - 2)^2 = (20 - 2)^2 = 18^2 = 324$

39. $3 \cdot 5 - 2^2 = 15 - 4 = 11$

41. $8 \div 4 \cdot 2 = 2 \cdot 2 = 4$

43. $36 + 4 \cdot 2 - 7 \cdot 6 = 36 + 8 - 42 = 44 - 42 = 2$

45. $-7, -3, -2, 0, 1, 4, 6$

47. Maximum: 5
Minimum: -6

49. -17

51. $|9| = 9$

53. $-|9| = -9$

55. $|12 - 8| = |4| = 4$

57. $-|8 + 12| = -|20| = -20$

59. $|-7| - |-3| = 7 - 3 = 4$

Chapter 1   The Language of Algebra

Exercises 1.1

1. The sum of c and d is written as $c + d$.

3. w plus z is written as $w + z$.

5. x increased by 2 is written as $x + 2$.

7. 10 more than y is written as $y + 10$.

9. a minus b is written as $a - b$.

11. b decreased by 7 is written as $b - 7$.

13. 6 less than r is written as $r - 6$.

15. w times z is written as wz.

17. The product of 5 and t is written as $5t$.

19. The product of 8, m, and n is written as $8mn$.

21. The product of 3 and the quantity p plus q is written as $3(p + q)$.

23. Twice the sum of x and y is written as $2(x + y)$.

25. The sum of twice x and y is written as $2x + y$.

27. Twice the difference of x and y is written as $2(x - y)$.

29. The quantity a plus b times the quantity a minus b is written as $(a + b)(a - b)$.

31. The product of m and 3 less than m is written as $m(m - 3)$.

33. x divided by 5 is written as $\dfrac{x}{5}$.

35. The quotient of a plus b, divided by 7 is written as $\dfrac{a+b}{7}$.

37. The difference of p and q, divided by 4 is written as $\dfrac{p-q}{4}$.

39. The sum of a and 3, divided by the difference of a and 3 is written as $\dfrac{a+3}{a-3}$.

41. 5 more than a number is written as $x + 5$.

43. 7 less than a number is written as $x - 7$.

45. 9 times a number is written as $9x$.

47. 6 more than 3 times a number is written as $3x + 6$.

49. Twice the sum of a number and 5 is written as $2(x + 5)$.

51. The product of 2 more than a number and 2 less than that same number is written as $(x + 2)(x - 2)$.

53. The quotient of a number and 7 is written as $\dfrac{x}{7}$.

55. The sum of a number and 5, divided by 8 is written as $\dfrac{x+5}{8}$.

57. 6 more than a number divided by 6 less than that same number is written as $\dfrac{x+6}{x-6}$.

59. Four times the length of a side (s) is written as $4s$.

61. The radius (r) squared times the height (h) times π is written as $r^2 h\pi$ or $\pi r^2 h$.

63. One-half the product of the height (h) and the sum of two unequal sides (b_1 and b_2) is written as $\dfrac{1}{2} h(b_1 + b_2)$.

65. $2(x + 5)$ is an expression. It means we multiply 2 by the sum of x and 5.

67. $4 + \div m$ is not an expression. The two operations in a row have no meaning.

69. $2b = 6$ is not an expression. The equals sign is not an operation sign.

71. $2a + 5b$ is an expression. It means we add 5 times b to 2 times a.

73. Let x be Earth's population 40 years ago. Then $2x$ represents Earth's population today.

75. The interest (*I*) equals the principal (*P*) times the rate (*r*) times the time (*t*) is written as $I = Prt$.

77. Writing exercise

Exercises 1.2

1. $5 + 9 = 9 + 5$ demonstrates the commutative property of addition.

3. $2 \cdot (3 \cdot 5) = (2 \cdot 3) \cdot 5$ demonstrates the associative property of multiplication.

5. $10 \cdot 5 = 5 \cdot 10$ demonstrates the commutative property of multiplication.

7. $8 + 12 = 12 + 8$ demonstrates the commutative property of addition.

9. $(5 \cdot 7) \cdot 2 = 5 \cdot (7 \cdot 2)$ demonstrates the associative property of multiplication.

11. $9 \cdot 8 = 8 \cdot 9$ demonstrates the commutative property of multiplication.

13. $2(3 + 5) = 2 \cdot 3 + 2 \cdot 5$ demonstrates the distributive property.

15. $5 + (7 + 8) = (5 + 7) + 8$ demonstrates the associative property of addition.

17. $(10 + 5) + 9 = 10 + (5 + 9)$ demonstrates the associative property of addition.

19. $7 \cdot (3 + 8) = 7 \cdot 3 + 7 \cdot 8$ demonstrates the distributive property.

21. $7 \cdot (3 + 4) = 7 \cdot 7 = 49$
$7 \cdot 3 + 7 \cdot 4 = 21 + 28 = 49$
Since $49 = 49$,
$7 \cdot (3 + 4) = 7 \cdot 3 + 7 \cdot 4$

23. $2 + (9 + 8) = 2 + 17 = 19$
$(2 + 9) + 8 = 11 + 8 = 19$
Since $19 = 19$,
$2 + (9 + 8) = (2 + 9) + 8$

25. $5 \cdot (6 \cdot 3) = 5 \cdot 18 = 90$
$(5 \cdot 6) \cdot 3 = 30 \cdot 3 = 90$
Since $90 = 90$,
$5 \cdot (6 \cdot 3) = (5 \cdot 6) \cdot 3$

27. $5 \cdot (2 + 8) = 5 \cdot 10 = 50$
$5 \cdot 2 + 5 \cdot 8 = 10 + 40 = 50$
Since $50 = 50$,
$5 \cdot (2 + 8) = 5 \cdot 2 + 5 \cdot 8$

29. $(3 + 12) + 8 = 15 + 8 = 23$
$3 + (12 + 8) = 3 + 20 = 23$
Since $23 = 23$,
$(3 + 12) + 8 = 3 + (12 + 8)$

31. $(4 \cdot 7) \cdot 2 = 28 \cdot 2 = 56$
$4 \cdot (7 \cdot 2) = 4 \cdot 14 = 56$
Since $56 = 56$,
$(4 \cdot 7) \cdot 2 = 4 \cdot (7 \cdot 2)$

33. $\dfrac{1}{2} \cdot (2 + 6) = \dfrac{1}{2} \cdot 8 = 4$
$\dfrac{1}{2} \cdot 2 + \dfrac{1}{2} \cdot 6 = 1 + 3 = 4$
Since $4 = 4$,
$\dfrac{1}{2} \cdot (2 + 6) = \dfrac{1}{2} \cdot 2 + \dfrac{1}{2} \cdot 6$

35. $\left(\dfrac{2}{3} + \dfrac{1}{6}\right) + \dfrac{1}{3} = \dfrac{5}{6} + \dfrac{1}{3} = \dfrac{7}{6}$
$\dfrac{2}{3} + \left(\dfrac{1}{6} + \dfrac{1}{3}\right) = \dfrac{2}{3} + \dfrac{3}{6} = \dfrac{7}{6}$

37. $(2.3 + 3.9) + 4.1 = 6.2 + 4.1 = 10.3$
$2.3 + (3.9 + 4.1) = 2.3 + 8.0 = 10.3$
Since $10.3 = 10.3$,
$(2.3 + 3.9) + 4.1 = 2.3 + (3.9 + 4.1)$

39. $\dfrac{1}{2} \cdot (2 \cdot 8) = \dfrac{1}{2} \cdot 16 = 8$
$\left(\dfrac{1}{2} \cdot 2\right) \cdot 8 = 1 \cdot 8 = 8$
Since $8 = 8$,
$\dfrac{1}{2} \cdot (2 \cdot 8) = \left(\dfrac{1}{2} \cdot 2\right) \cdot 8$

41. $\left(\dfrac{3}{5} \cdot \dfrac{5}{6}\right) \cdot \dfrac{4}{3} = \dfrac{1}{2} \cdot \dfrac{4}{3} = \dfrac{2}{3}$
$\dfrac{3}{5} \cdot \left(\dfrac{5}{6} \cdot \dfrac{4}{3}\right) = \dfrac{3}{5} \cdot \dfrac{10}{9} = \dfrac{2}{3}$
Since $\dfrac{2}{3} = \dfrac{2}{3}$,
$\left(\dfrac{3}{5} \cdot \dfrac{5}{6}\right) \cdot \dfrac{4}{3} = \dfrac{3}{5} \cdot \left(\dfrac{5}{6} \cdot \dfrac{4}{3}\right)$

43. $2.5 \cdot (4 \cdot 5) = 2.5 \cdot 20 = 50$
$(2.5 \cdot 4) \cdot 5 = 10 \cdot 5 = 50$
Since $50 = 50$,
$2.5 \cdot (4.5) = (2.5 \cdot 4) \cdot 5$

45. $2(3 + 5) = 2 \cdot 3 + 2 \cdot 5$
$\qquad = 6 + 10$
$\qquad = 16$

47. $3(x + 5) = 3 \cdot x + 3 \cdot 5$
$\qquad = 3x + 15$

49. $4(w + v) = 4 \cdot w + 4 \cdot v$
$\qquad = 4w + 4v$

51. $2(3x + 5) = 2 \cdot 3x + 2 \cdot 5$
$\qquad = 6x + 10$

53. $\dfrac{1}{3} \cdot (15 + 9) = \dfrac{1}{3} \cdot 15 + \dfrac{1}{3} \cdot 9$
$\qquad\qquad = 5 + 3$
$\qquad\qquad = 8$

55. $5 + 7 = 7 + 5$ by the commutative property of addition.

57. $(8)(3) = (3)(8)$ by the commutative property of multiplication.

59. $7(2 + 5) = 7 \cdot 2 + 7 \cdot 5$ by the distributive property.

61. $3 + 7 = 7 + 3$ by the commutative property of addition.

63. $5 \cdot (3 \cdot 2) = (5 \cdot 3) \cdot 2$ by the associative property of multiplication.

65. $2 \cdot 4 + 2 \cdot 5 = 2 \cdot (4 + 5)$ by the distributive property.

67. $8 - 5 = 3$
$5 - 8 = -3$
Since 3 is not equal to -3,
subtraction is not commutative.

69. $(12 - 8) - 4 = 4 - 4 = 0$
$12 - (8 - 4) = 12 - 4 = 8$
Since 0 is not equal to 8, subtraction is not associative.

71. $3(6 - 2) = 3(4) = 12$
$3 \cdot 6 - 3 \cdot 2 = 18 - 6 = 12$

Since $12 = 12$, multiplication is distributive over subtraction.

73. a. $5 \cdot (3 + 4) = 5 \cdot 3 + 5 \cdot 4$ by the distributive property.

b. $5 \cdot (3 + 4) = 5 \cdot (4 + 3)$ by the commutative property of addition.

c. $5 \cdot (3 + 5) = (3 + 4) \cdot 5$ by the commutative property of multiplication.

75. $5 + (6 + 7) = (5 + 6) + 7$ demonstrates the associative property of addition.

77. $4 \cdot (3 + 2) = 4 \cdot (2 + 3)$ demonstrates the commutative property of addition.

Exercises 1.3

1. $3 + 6 = 9$

3. $11 + 5 = 16$

5. $\dfrac{3}{4} + \dfrac{5}{4} = \dfrac{8}{4} = 2$

7. $\dfrac{1}{2} + \dfrac{4}{5} = \dfrac{5}{10} + \dfrac{8}{10} = \dfrac{13}{10}$

9. $(-2) + (-3) = -5$

11. $\left(-\dfrac{3}{5}\right) + \left(-\dfrac{7}{5}\right) = -\dfrac{10}{5} = -2$

13. $\left(-\dfrac{1}{2}\right) + \left(-\dfrac{3}{8}\right) = \left(-\dfrac{4}{8}\right) + \left(-\dfrac{3}{8}\right) = -\dfrac{7}{8}$

15. $(-1.6) + (-2.3) = -3.9$

17. $9 + (-3) = 6$

19. $\dfrac{3}{4} \div \left(-\dfrac{1}{2}\right) = \dfrac{3}{4} + \left(-\dfrac{2}{4}\right) = \dfrac{1}{4}$

21. $\left(-\dfrac{4}{5}\right) + \dfrac{9}{20} = \left(-\dfrac{16}{20}\right) + \dfrac{9}{20} = -\dfrac{7}{20}$

23. $-11.4 + 13.4 = 2$

25. $-3.6 + 7.6 = 4$

27. $-9 + 0 = -9$

29. $7 + (-7) = 0$

31. $-4.5 + 4.5 = 0$

33. $7 + (-9) + (-5) + 6 = 13 + (-14) = -1$

35. $7 + (-3) + 5 + (-11) = 12 + (-14) = -2$

37. $-\dfrac{3}{2} + \left(-\dfrac{7}{4}\right) + \dfrac{1}{4} = -\dfrac{6}{4} + \left(-\dfrac{7}{4}\right) + \dfrac{1}{4} = -\dfrac{12}{4} = -3$

39. $2.3 + (-5.4) + (-2.9) = -6$

41. $21 - 13 = 21 + (-13) = 8$

43. $82 - 45 = 82 + (-45) = 37$

45. $\dfrac{15}{7} - \dfrac{8}{7} = \dfrac{15}{7} + \left(-\dfrac{8}{7}\right) = \dfrac{7}{7} = 1$

47. $7.9 - 5.4 = 7.9 + (-5.4) = 2.5$

49. $8 - 10 = 8 + (-10) = -2$

51. $24 - 45 = 24 + (-45) = -21$

53. $\dfrac{7}{6} - \dfrac{19}{6} = \dfrac{7}{6} + \left(-\dfrac{19}{6}\right) = -\dfrac{12}{6} = -2$

55. $7.8 - 11.6 = 7.8 + (-11.6) = -3.8$

57. $-5 - 3 = -5 + (-3) = -8$

59. $-9 - 14 = -9 + (-14) = -23$

61. $-\dfrac{2}{5} - \dfrac{7}{10} = -\dfrac{4}{10} + \left(-\dfrac{7}{10}\right) = -\dfrac{11}{10}$

63. $-3.4 - 4.7 = -3.4 + (-4.7) = -8.1$

65. $5 - (-11) = 5 + 11 = 16$

67. $7 - (-12) = 7 + 12 = 19$

69. $\dfrac{3}{4} - \left(-\dfrac{3}{2}\right) = \dfrac{3}{4} + \dfrac{3}{2} = \dfrac{3}{4} + \dfrac{6}{4} = \dfrac{9}{4}$

71. $\dfrac{6}{7} - \left(-\dfrac{5}{14}\right) = \dfrac{6}{7} + \dfrac{5}{14} = \dfrac{12}{14} + \dfrac{5}{14} = \dfrac{17}{14}$

73. $8.3 - (-5.7) = 8.3 + 5.7 = 14$

75. $8.9 - (-11.7) = 8.9 + 11.7 = 20.6$

77. $-36 - (-24) = -36 + 24 = -12$

79. $-19 - (-27) = -19 + 27 = 8$

81. $\left(-\dfrac{3}{4}\right) - \left(-\dfrac{11}{4}\right) = -\dfrac{3}{4} + \dfrac{11}{4} = \dfrac{8}{4} = 2$

83. $-12.7 - (-5.7) = -12.7 + 5.7 = -7$

85. $-6.9 - (-10.1) = -6.9 + 10.1 = 3.2$

87. $-4.1967 - 5.2943 = -9.491$

89. $-4.1623 - (-3.1468) = -1.0155$

91. $-6.3267 + 8.6789 = 2.3522$

93. 1, 3, 5, 7, 9
The median is the element that is exactly in the middle of the set written in ascending order.
The median for this set is 5.

95. 8, 7, 2, 25, 5, 13, 3
First, rewrite the set in ascending order.
2, 3, 5, 7, 8, 13, 25
The median is the element that is exactly in the middle. The median for this set is 7.

97. 2, 7, 9, 15, 24
The set is already written in ascending order. The range is the difference between the maximum, 24, and the minimum, 2.
$24 - 2 = 22$
The range is 22.

99. −4, −3, 2, 7, 9
The set is already written in ascending order. The range is the difference between the maximum, 9, and the minimum, −4.
$9 - (-4) = 9 + 4 = 13$
The range is 13.

101. $\dfrac{7}{8}$, 2, $-\dfrac{1}{2}$, −8, $\dfrac{3}{4}$
First, rewrite the set in ascending order.
$-8, -\dfrac{1}{2}, \dfrac{3}{4}, \dfrac{7}{8}, 2$
The maximum is 2. The minimum is −8.
The range is $2 - (-8) = 2 + 8 = 10$.

103. 3, 2, −5, 6, −3
First, rewrite the set in ascending order.
−5, −3, 2, 3, 6
The maximum is 6. The minimum is −5.
The range is 6 − (−5) = 6 + 5 = 11.

105. 100 + (−23) + 51 = 128
His new balance is $128.

107. 23 + (−5) + 15 + (−10) = 23
His net yardage change is 23 yards gained.

109. 82 + (−12) = 70
The temperature was 70° at 4:00 P.M.

111. −72 + (−23.50) = −95.5
His checking account was overdrawn by
$95.50.

113. −750 + (−425) = −1175
The total decrease in enrollment was
1175 students.

115. 87, 71, 95, 81, 90
First, rewrite the set in ascending order.
71, 81, 87, 90, 95
The maximum is 95. The minimum is 71.
The range is 95 − 71 = 24.

117. Writing exercise

119. −1, 3, 5, −2, 4, 12, 10
First, rewrite the set in ascending order.
−2, −1, 3, 4, 5, 10, 12
The maximum is 12. The minimum is −2.
The range is 12 − (−2) = 12 + 2 = 14 or 14°F.

121. Writing exercise

Exercises 1.4

1. $4 \cdot 10 = 40$

3. $(5)(-12) = -60$

5. $(-8)(9) = -72$

7. $(4)\left(-\dfrac{3}{2}\right) = -6$

9. $\left(-\dfrac{1}{4}\right)(8) = -2$

11. $(3.25)(-4) = -13$

13. $(-8)(-7) = 56$

15. $(-5)(-12) = 60$

17. $(-9)\left(-\dfrac{2}{3}\right) = 6$

19. $(-1.25)(-12) = 15$

21. $(0)(-18) = 0$

23. $(15)(0) = 0$

25. $\left(-\dfrac{11}{12}\right)(0) = 0$

27. $(-3.57)(0) = 0$

29. $\left(-\dfrac{3}{2}\right)\left(-\dfrac{2}{3}\right) = 1$

31. $\left(\dfrac{4}{7}\right)\left(-\dfrac{7}{4}\right) = -1$

33. $\dfrac{-20}{-4} = 5$

35. $\dfrac{48}{6} = 8$

37. $\dfrac{50}{-5} = -10$

39. $\dfrac{-52}{4} = -13$

41. $\dfrac{-75}{-3} = 25$

43. $\dfrac{0}{-8} = 0$

45. $\dfrac{-9}{-1} = 9$

47. $\dfrac{-96}{-8} = 12$

49. $\dfrac{18}{0}$ is undefined.

51. $\dfrac{-17}{1} = -17$

53. $\dfrac{-144}{-16} = 9$

55. $\dfrac{-29.4}{4.9} = -6$

57. $\dfrac{-8}{32} = -\dfrac{1}{4}$

59. $\dfrac{24}{-16} = -\dfrac{3}{2}$

61. $\dfrac{-28}{-42} = \dfrac{2}{3}$

63. $\dfrac{(-6)(-3)}{2} = \dfrac{18}{2} = 9$

65. $\dfrac{(-8)(2)}{-4} = \dfrac{-16}{-4} = 4$

67. $\dfrac{24}{-4-8} = \dfrac{24}{-12} = -2$

69. $\dfrac{-12-12}{-3} = \dfrac{-24}{-3} = 8$

71. $\dfrac{55-19}{-12-6} = \dfrac{36}{-18} = -2$

73. $\dfrac{7-5}{2-2} = \dfrac{2}{0}$ is undefined.

75. $5(7-2) = 5(5) = 25$

77. $2(5-8) = 2(-3) = -6$

79. $-3(9-7) = -3(2) = -6$

81. $-3(-2-5) = -3(-7) = 21$

83. $(-2)(3) - 5 = -6 - 5 = -11$

85. $4(-7) - 5 = -28 - 5 = -33$

87. $(-5)(-2) - 12 = 10 - 12 = -2$

89. $(3)(-7) + 20 = -21 + 20 = -1$

91. $-4 + (-3)(6) = -4 + (-18) = -22$

93. $7 - (-4)(-2) = 7 - 8 = -1$

95. $(-7)^2 - 17 = 49 - 17 = 32$

97. $(-5)^2 + 18 = 25 + 18 = 43$

99. $-6^2 - 4 = -36 - 4 = -40$

101. $(-4)^2 - (-2)(-5) = 16 - (-2)(-5)$
$ = 16 - 10$
$ = 6$

103. $(-8)^2 - 5^2 = 64 - 25$
$ = 39$

105. $(-6)^2 - (-3)^2 = 36 - 9 = 27$

107. $-8^2 - 5^2 = -64 - 25 = -89$

109. $-8^2 - (-5)^2 = -64 - 25 = -89$

111. $23 \cdot 11 = 253$
I scored a total of 253 points.

113. $335 \cdot 1.25 = 43.75$
I made $43.75.

115. $-6 - (2 \cdot 8) = -22$
The temperature is $-22°$F.

117. $125 - (9 \cdot 9) = 44$
He had $44.

119. $\dfrac{16{,}232 - 20{,}000}{3} = \dfrac{-3768}{3} = -1256$
Each person lost $1256.

121. $84 \div \dfrac{2}{3} = 126$
He can fill 126 test tubes.

123. $\dfrac{-8}{-4+2} = 4$

125. $\dfrac{-10+4}{-7+10} = -2$

Exercises 1.5

For exercises 1–41, $a = -2$, $b = 5$ and $c = -4$, and $d = 6$.

1. $3c - 2b = 3(-4) - 2(5)$
$\qquad = -12 - 10$
$\qquad = -22$

3. $8b + 2c = 8(5) + 2(-4)$
$\qquad = 40 + (-8)$
$\qquad = 32$

5. $-b^2 + b = -5^2 + 5$
$\qquad = -25 + 5$
$\qquad = -20$

7. $3a^2 = 3(-2)^2$
$\qquad = 3(4)$
$\qquad = 12$

9. $c^2 - 2d = (-4)^2 - 2(6)$
$\qquad = 16 - 12$
$\qquad = 4$

11. $2a^2 + 3b^2 = 2(-2)^2 + 3(5)^2$
$\qquad = 2(4) + 3(25)$
$\qquad = 8 + 75$
$\qquad = 83$

13. $2(a + b) = 2(-2 + 5)$
$\qquad = 2(3)$
$\qquad = 6$

15. $4(2a - d) = 4[2(-2) - 6]$
$\qquad = 4(-4 - 6)$
$\qquad = 4(-10)$
$\qquad = -40$

17. $a(b + 3c) = -2[5 + 3(-4)]$
$\qquad = -2(5 - 12)$
$\qquad = -2(-7)$
$\qquad = 14$

19. $\dfrac{6d}{c} = \dfrac{6 \cdot 6}{-4}$
$\qquad = \dfrac{36}{-4}$
$\qquad = -9$

21. $\dfrac{3d + 2c}{b} = \dfrac{3(6) + 2(-4)}{5}$
$\qquad = \dfrac{18 + (-8)}{5}$
$\qquad = \dfrac{10}{5}$
$\qquad = 2$

23. $\dfrac{2b - 3a}{c + 2d} = \dfrac{2(5) - 3(-2)}{-4 + 2(6)}$
$\qquad = \dfrac{10 + 6}{-4 + 12}$
$\qquad = \dfrac{16}{8}$
$\qquad = 2$

25. $d^2 - b^2 = 6^2 - 5^2$
$\qquad = 36 - 25$
$\qquad = 11$

27. $(d - b)^2 = (6 - 5)^2$
$\qquad = 1^2$
$\qquad = 1$

29. $(d - b)(d + b) = (6 - 5)(6 + 5) = (1)(11) = 11$

31. $d^3 - b^3 = (6)^3 - (5)^3 = 216 - 125 = 91$

33. $(d - b)^3 = (6 - 5)^3 = 1^3 = 1$

35. $(d - b)(d^2 + db + b^2) = (6 - 5)[6^2 + (6)(5) + 5^2]$
$\qquad = (1)(36 + 30 + 25)$
$\qquad = (1)(91)$
$\qquad = 91$

37. $b^2 + a^2 = (5)^2 + (-2)^2 = 225 + 4 = 29$

39. $(b + a)^2 = [5 + (-2)]^2 = 3^2 = 9$

41. $a^2 + 2ad + d^2 = (-2)^2 + 2(-2)(6) + (6)^2$
$\qquad = 4 - 24 + 36$
$\qquad = 16$

For exercises 43–49, $x = -2.34$, $y = -3.14$, and $z = 4.12$.

43. $x + yz = -2.34 + (-3.14)(4.12) = -15.3$

45. $x^2 - z^2 = (-2.34)^2 - (4.12)^2 = -11.5$

47. $\dfrac{xy}{z-x} = \dfrac{(-2.34)(-3.14)}{4.12-(-2.34)} = 1.1$

49. $\dfrac{2x+y}{2x+z} = \dfrac{2(-2.34)+(-3.14)}{2(-2.34)+4.12} = 14.0$

51. $\sum x = 1+2+3+7+8+9+11 = 41$

53. $\sum x = -5+(-3)+(-1)+2+3+4+8 = 8$

55. $\sum x = 3+2+(-1)+(-4)+(-3)+8+6 = 11$

57. $\sum x = -\dfrac{1}{2}+\left(-\dfrac{3}{4}\right)+2+3+\dfrac{1}{4}+\dfrac{3}{2}+(-1) = \dfrac{9}{2}$

59. $\sum x = -2.5+(-3.2)+2.6+(-1)+2+4+(-3)$
$= -1.1$

61. $x-7 = 22-7 = 15$
$2y+5 = 2(5)+5 = 15$
The statement is true.

63. $2(x+y) = 2[-4+(-2)] = 2(-6) = -12$
$2x+y = 2(-4)+(-2) = -10$
The statement is false.

65. $R_T = \dfrac{R_1 R_2}{(R_1+R_2)}$
$= \dfrac{(6 \cdot 10)}{(6+10)} = 3.75$
The total resistance is 3.75 ohms.

67. $P = 2L+2W = 2(10)+2(5) = 30$
The perimeter is 30 inches.

69. $P = \dfrac{I}{RT} = \dfrac{150}{(0.04)(2)} = 1875$
The principal is $1875.

71. $F = \dfrac{9}{5}C+32 = \dfrac{9}{5}(-10)+32 = 14$
The temperature is 14° F.

73. Writing exercise

75. Writing exercise

Exercises 1.6

1. $5a+2$ has two terms: $5a$ and 2.

3. $4x^3$ has one term: $4x^3$.

5. $3x^2+3x-7$ or $3x^2+3x+(-7)$ has three terms: $3x^2$, $3x$, and -7.

7. In the group of terms $5ab$, $3b$, $3a$, $4ab$, the like terms are $5ab$ and $4ab$.

9. In the group of terms $4xy^2$, $2x^2y$, $5x^2$, $-3x^2y$, $5y$, $6x^2y$, the like terms are $2x^2y$, $-3x^2y$, and $6x^2y$.

11. $3m+7m = (3+7)m = 10m$

13. $7b^3+10b^3 = (7+10)b^3 = 17b^3$

15. $21xyz+7xyz = (21+7)xyz = 28xyz$

17. $9z^2-3z^2 = 9z^2+(-3z^2) = 6z^2$

19. $5a^3-5a^3 = 5a^3+(-5a^3) = 0$

21. $19n^2-18n^2 = 19n^2+(-18n^2) = 1n^2 = n^2$

23. $21p^2q-6p^2q = 21p^2q+(-6p^2q) = 15p^2q$

25. $10x^2-7x^2+3x^2 = 10x^2+(-7x^2)+3x^2$
$= (10+(-7)+3)x^2$
$= 6x^2$

27. $9a-7a+4b = 9a+(-7a)+4b$
$= (9+(-7))a+4b$
$= 2a+4b$

29. $7x+5y-4x-4y = 7x+(-4x)+5y+(-4y)$
$= (7+(-4))x+(5+(-4))y$
$= 3x+1y$
$= 3x+y$

31. $4a+7b+3-2a+3b-2$
$= 4a+(-2a)+7b+3b+3+(-2)$
$= (4+(-2))a+(7+3)b+3+(-2)$
$= 2a+10b+1$

33. $\dfrac{2}{3}m+3+\dfrac{4}{3}m = \left(\dfrac{2}{3}+\dfrac{4}{3}\right)m+3$
$= \dfrac{6}{3}m+3$
$= 2m+3$

35. $\dfrac{13}{5}x + 2 - \dfrac{3}{5}x + 5 = \left(\dfrac{13}{5} - \dfrac{3}{5}\right)x + 2 + 5$

$\qquad\qquad\qquad\quad = \dfrac{10}{5}x + 7$

$\qquad\qquad\qquad\quad = 2x + 7$

37. $2.3a + 7 + 4.7a + 3 = (2.3 + 4.7)a + 7 + 3$

$\qquad\qquad\qquad\qquad\; = 7a + 10$

39. $5a^4 + 8a^4 = (5 + 8)a^4$

$\qquad\qquad\quad = 13a^4$

41. $15a^3 - 12a^3 = 15a^3 + (-12a^3)$

$\qquad\qquad\qquad = 3a^3$

43. $(8x + 3x) - 4x = 11x - 4x$

$\qquad\qquad\qquad = 7x$

45. $(9mn^2 + 5mn^2) - 3mn^2 = 14mn^2 - 3mn^2$

$\qquad\qquad\qquad\qquad\qquad = 11mn^2$

47. $2(3x + 2) + 4 = 6x + 4 + 4$

$\qquad\qquad\qquad = 6x + 8$

49. $5(6a - 2) + 12a = 30a - 10 + 12a$

$\qquad\qquad\qquad\quad = 42a - 10$

51. $4s + 2(s + 4) + 4 = 4s + 2s + 8 + 4$

$\qquad\qquad\qquad\qquad = 6s + 12$

53. Writing exercise

55. Writing exercise

57. Group exercise

Exercises 1.7

1. $x^5 \cdot x^7 = x^{5+7} = x^{12}$

3. $5 \cdot 5^5 = 5^1 \cdot 5^5 = 5^{1+5} = 5^6$

5. $a^9 \cdot a = a^9 \cdot a^1 = a^{9+1} = a^{10}$

7. $z^{10} \cdot z^3 = z^{10+3} = z^{13}$

9. $p^5 \cdot p^7 = p^{5+7} = p^{12}$

11. $x^3 y \cdot x^2 y^4 = x^{3+2} y^{1+4} = x^5 y^5$

13. $w^5 \cdot w^2 \cdot w = w^{5+2+1} = w^8$

15. $m^3 \cdot m^2 \cdot m^4 = m^{3+2+4} = m^9$

17. $a^3 b \cdot a^2 b^2 \cdot ab^3 = a^{3+2+1} b^{1+2+3} = a^6 b^6$

19. $p^2 q \cdot p^3 q^5 \cdot pq^4 = p^{2+3+1} q^{1+5+4}$

$\qquad\qquad\qquad\qquad = p^6 q^{10}$

21. $3a^6 \cdot 2a^3 = (3 \cdot 2)(a^6 \cdot a^3)$

$\qquad\qquad\quad = 6a^9$

23. $x^2 \cdot 3x^5 = (1 \cdot 3)(x^2 \cdot x^5)$

$\qquad\qquad\quad = 3x^7$

25. $5m^3 n^2 \cdot 4mn^3 = (5 \cdot 4)(m^3 \cdot m)(n^2 \cdot n^3)$

$\qquad\qquad\qquad\quad = 20m^4 n^5$

27. $6x^3 y \cdot 9xy^5 = (6 \cdot 9)(x^3 \cdot x)(y \cdot y^5)$

$\qquad\qquad\qquad = 54x^4 y^6$

29. $2a^2 \cdot a^3 \cdot 3a^7 = (2 \cdot 1 \cdot 3)(a^2 \cdot a^3 \cdot a^7)$

$\qquad\qquad\qquad\quad = 6a^{12}$

31. $3c^2 d \cdot 4cd^3 \cdot 2c^5 d = (3 \cdot 4 \cdot 2)(c^2 \cdot c \cdot c^5)(d \cdot d^3 \cdot d)$

$\qquad\qquad\qquad\qquad\quad = 24c^8 d^5$

33. $5m^2 \cdot m^3 \cdot 2m \cdot 3m^4$

$\quad = (5 \cdot 1 \cdot 2 \cdot 3)(m^2 \cdot m^3 \cdot m \cdot m^4)$

$\quad = 30m^{10}$

35. $2r^3 s \cdot rs^2 \cdot 3r^2 s \cdot 5rs$

$\quad = (2 \cdot 1 \cdot 3 \cdot 5)(r^3 \cdot r \cdot r^2 \cdot r)(s \cdot s^2 \cdot s \cdot s)$

$\quad = 30r^7 s^5$

37. $\dfrac{a^9}{a^6} = a^{9-6} = a^3$

39. $\dfrac{y^{10}}{y^4} = y^{10-4} = y^6$

41. $\dfrac{p^{15}}{p^{10}} = p^{15-10} = p^5$

43. $\dfrac{x^5 y^3}{x^2 y^2} = x^{5-2} \cdot y^{3-2} = x^3 y$

45. $\dfrac{6m^3}{3m} = 2m^{3-1} = 2m^2$

47. $\dfrac{24a^7}{6a^4} = 4a^{7-4} = 4a^3$

49. $\dfrac{26m^8 n}{13m^6} = 2m^{8-6} \cdot n = 2m^2 n$

51. $\dfrac{28w^3 z^5}{7wz} = 4w^{3-1} \cdot z^{5-1}$
$\qquad = 4w^2 z^4$

53. $\dfrac{18x^3 y^4 z^5}{9xy^2 z^2} = 2x^{3-1} \cdot y^{4-2} \cdot z^{5-2}$
$\qquad = 2x^2 y^2 z^3$

55. $2a^3 b \cdot 3a^2 b = (2 \cdot 3)(a^3 \cdot a^2)(b \cdot b)$
$\qquad = 6a^5 b^2$

57. $2a^3 b + 3a^2 b$ cannot be simplified.
The bases are not the same.

59. $2x^2 y^3 \cdot 3x^2 y^3 = (2 \cdot 3)(x^2 \cdot x^2)(y^3 \cdot y^3)$
$\qquad = 6x^4 y^6$

61. $2x^2 y^3 + 3x^2 y^3 = (2 + 3)x^2 y^3$
$\qquad = 5x^2 y^3$

63. $\dfrac{8a^2 b \cdot 6a^2 b}{2ab} = \dfrac{(8 \cdot 6)a^{2+2}b^{1+1}}{2ab}$
$\qquad = \dfrac{48a^4 b^2}{2ab}$
$\qquad = 24a^{4-1} \cdot b^{2-1}$
$\qquad = 24a^3 b$

65. $\dfrac{8a^2 b + 6a^2 b}{2ab} = \dfrac{(8 + 6)a^2 b}{2ab}$
$\qquad = \dfrac{14a^2 b}{2ab}$
$\qquad = 7a^{2-1}b^{1-1}$
$\qquad = 7a^1 b^0$
$\qquad = 7a$

67. Writing exercise

69. Writing exercise

Summary Exercises for Chapter 1

1. 5 more than y is written as $y + 5$.

3. The product of 8 and a is written as $8a$.

5. 5 times the product of m and n is written as $5mn$.

7. 3 more than the product of 17 and x is written as $17x + 3$.

9. $4(x + 3)$ is an expression. It means we multiply 4 by the sum of x and 3.

11. $y + 5 = 9$ is not an expression. The equals sign is not an operation sign.

13. $5 + (7 + 12) = (5 + 7) + 12$ demonstrates the associative property of addition.

15. $4 \cdot (5 \cdot 3) = (4 \cdot 5) \cdot 3$ demonstrates the associative property of multiplication.

17. $8(5 + 4) = 8(9) = 72$
$8 \cdot 5 + 8 \cdot 4 = 40 + 32 = 72$
Since $72 = 72$,
$8(5 + 4) = 8 \cdot 5 + 8 \cdot 4$

19. $(7 + 9) + 4 = 16 + 4 = 20$
$7 + (9 + 4) = 7 + 13 = 20$
Since $20 = 20$,
$(7 + 9) + 4 = 7 + (9 + 4)$

21. $(8 \cdot 2) \cdot 5 = 16 \cdot 5 = 80$
$8(2 \cdot 5) = 8(10) = 80$
Since $80 = 80$,
$(8 \cdot 2) \cdot 5 = 8(2 \cdot 5)$

23. $3(7 + 4) = 3 \cdot 7 + 3 \cdot 4$

25. $4(w + v) = 4 \cdot w + 4 \cdot v$
$\qquad = 4w + 4v$

27. $3(5a + 2) = 3 \cdot 5a + 3 \cdot 2$

29. $-3 + (-8) = -11$

31. $6 + (-6) = 0$

33. $-18 + 0 = -18$

35. $5.7 + (-9.7) = -4$

37. $8 - 13 = 8 + (-13) = -5$

39. $10 - (-7) = 10 + 7 = 17$

41. $-9 - (-9) = -9 + 9 = 0$

43. $-\dfrac{5}{4} - \left(-\dfrac{17}{4}\right) = -\dfrac{5}{4} + \dfrac{17}{4} = \dfrac{12}{4} = 3$

45. 2, 4, 9, 10, 15
The set is already written in ascending order.
The median is the element that is exactly in the middle of the set.
The median for this set is 9.

47. –3, –8, 4, 1, 6
First, rewrite the set in ascending order.
–8, –3, 1, 4, 6
The median is the element that is exactly in the middle of the set.
The median for this set is 1.

49. 2, 4, 1, 8, 6, 7
First, rewrite the set in ascending order.
1, 2, 4, 6, 7, 8
The median is the sum of the middle two numbers, divided by 2.
The median for this set is $\dfrac{4+6}{2} = \dfrac{10}{2} = 5$.

51. 3, 5, 1, 8, 9
First, rewrite the set in ascending order.
1, 3, 5, 8, 9
The maximum is 9. The minimum is 1.
The range is $9 - 1 = 8$.

53. –5, 2, –1, 3, 8
First, rewrite the set in ascending order.
–5, –1, 2, 3, 8
The maximum is 8. The minimum is –5.
The range is $8 - (-5) = 8 + 5 = 13$.

55. $(10)(-7) = -70$

57. $(-3)(-15) = 45$

59. $(0)(-8) = 0$

61. $(-4)\left(\dfrac{3}{8}\right) = -\dfrac{3}{2}$

63. $\dfrac{80}{16} = 5$

65. $\dfrac{-81}{-9} = 9$

67. $\dfrac{32}{-8} = -4$

69. $\dfrac{-8+6}{-8-(-10)} = \dfrac{-2}{-8+10}$
$= \dfrac{-2}{2}$
$= -1$

71. $\dfrac{25-4}{-5-(-2)} = \dfrac{25+(-4)}{-5+2}$
$= \dfrac{21}{-3}$
$= -7$

73. $(18-3)\cdot 5 = 15\cdot 5$
$= 75$

75. $(5\cdot 4)^2 = 20^2 = 400$

77. $5(3^2 - 4) = 5(9 - 4) = 5(5) = 25$

79. $5\cdot 4 - 2^2 = 5\cdot 4 - 4 = 20 - 4 = 16$

81. $3(5-2)^2 = 3(3)^2 = 3(9) = 27$

83. $(3\cdot 5 - 2)^2 = (15-2)^2 = 13^2 = 169$

For exercises 85–95, $x = -3$, $y = 6$, $z = -4$, and $w = 2$.

85. $5y - 4z = 5(6) - 4(-4)$
$= 30 + 16$
$= 46$

87. $5z^2 = 5(-4)^2$
$= 5(16)$
$= 80$

89. $3x^3 = 3(-3)^3$
$= 3(-27)$
$= -81$

91. $\dfrac{6z}{2w} = \dfrac{6(-4)}{2(2)}$
$= \dfrac{-24}{4}$
$= -6$

93. $\dfrac{3x-y}{w-x} = \dfrac{3(-3)-6}{2-(-3)}$

$\phantom{\dfrac{3x-y}{w-x}} = \dfrac{-9-6}{2+3}$

$\phantom{\dfrac{3x-y}{w-x}} = \dfrac{-15}{5}$

$\phantom{\dfrac{3x-y}{w-x}} = -3$

95. $\dfrac{y(x-w)^2}{x^2-2xw+w^2} = \dfrac{6(-3-2)^2}{(-3)^2-2(-3)(2)+2^2}$

$\phantom{\dfrac{y(x-w)^2}{x^2-2xw+w^2}} = \dfrac{6(-5)^2}{9-(-12)+4}$

$\phantom{\dfrac{y(x-w)^2}{x^2-2xw+w^2}} = \dfrac{6(25)}{9+12+4}$

$\phantom{\dfrac{y(x-w)^2}{x^2-2xw+w^2}} = \dfrac{150}{25}$

$\phantom{\dfrac{y(x-w)^2}{x^2-2xw+w^2}} = 6$

97. $5x^2-7x+3$ or $5x^2+(-7x)+3$ has three terms: $5x^2$, $-7x$, and 3.

99. In the group of terms
$4ab^2$, $3b^2-5a$, ab^2, $7a^2$, $-3ab^2$, $4a^2b$, the like terms are $4ab^2$, ab^2, and $-3ab^2$.

101. $2x+5x = (2+5)x$

$ = 7x$

103. $6c-3c = 6c+(-3c)$

$ = 3c$

105. $5ab^2+2ab^2 = (5+2)ab^2$

$ = 7ab^2$

107. $6x-2x+5y-3x = 6x+(-2x)+(-3x)+5y$

$ = (6+(-2)+(-3))x+5y$

$ = 1x+5y$

$ = x+5y$

109. $3a^3+5a^2+4a-2a^3-3a^2-a$

$= 3a^3+(-2a^3)+5a^2+(-3a^2)+4a+(-1a)$

$= a^3+2a^2+3a$

111. $15x^2-(3x^2+5x^2) = 15x^2-8x^2$

$ = 7x^2$

113. $\dfrac{a^5}{a^4} = a^{5-4} = a^1 = a$

115. $\dfrac{m^2 \cdot m^3 \cdot m^4}{m^5} = \dfrac{m^{2+3+4}}{m^5}$

$\phantom{\dfrac{m^2 \cdot m^3 \cdot m^4}{m^5}} = \dfrac{m^9}{m^5}$

$\phantom{\dfrac{m^2 \cdot m^3 \cdot m^4}{m^5}} = m^{9-5}$

$\phantom{\dfrac{m^2 \cdot m^3 \cdot m^4}{m^5}} = m^4$

117. $\dfrac{24x^{17}}{8x^{13}} = 3x^{17-13}$

$\phantom{\dfrac{24x^{17}}{8x^{13}}} = 3x^4$

119. $\dfrac{108x^9y^4}{9xy^4} = 12x^{9-1} \cdot y^{4-4}$

$\phantom{\dfrac{108x^9y^4}{9xy^4}} = 12x^8y^0$

$\phantom{\dfrac{108x^9y^4}{9xy^4}} = 12x^8$

121. $\dfrac{52a^5b^3c^5}{13a^4c} = 4a^{5-4} \cdot b^3 \cdot c^{5-1}$

$\phantom{\dfrac{52a^5b^3c^5}{13a^4c}} = 4ab^3c^4$

123. $(3x)^2(4xy) = (9x^2)(4xy)$

$ = (9 \cdot 4)(x^2 \cdot x)(y)$

$ = 36x^3y$

125. $(-2x^3y^3)(-5xy) = [-2 \cdot (-5)](x^3 \cdot x)(y^3 \cdot y)$

$ = 10x^4y^4$

127. Subtract x from 23.
$23-x$

129. Add 5 years to Angela's age, x.
$x+5$

131. Add 4 to the width. Let x = the width.
$x+4$

133. x = length of one piece
$25-x$ = length of the other piece

Chapter 2   Equations and Inequalities

Exercises 2.1

1. $x+4=9$
$5+4 \overset{?}{=} 9$
$9=9$
5 is a solution.

3. $x-15=6$
$-21-15 \overset{?}{=} 6$
$-36 \neq 6$
-21 is not a solution.

5. $5-x=2$
$5-4 \overset{?}{=} 2$
$1 \neq 2$
4 is not a solution.

7. $4-x=6$
$4-(-2) \overset{?}{=} 6$
$4+2 \overset{?}{=} 6$
$6=6$
-2 is a solution.

9. $3x+4=13$
$3(8)+4 \overset{?}{=} 13$
$24+4 \overset{?}{=} 13$
$28 \neq 13$
8 is not a solution.

11. $4x-5=7$
$4(2)-5 \overset{?}{=} 7$
$8-5 \overset{?}{=} 7$
$3 \neq 7$
2 is not a solution.

13. $5-2x=7$
$5-2(-1) \overset{?}{=} 7$
$5+2 \overset{?}{=} 7$
$7=7$
-1 is a solution.

15. $4x-5=2x+3$
$4(4)-5 \overset{?}{=} 2(4)+3$
$16-5 \overset{?}{=} 8+3$
$11=11$
4 is a solution.

17. $x+3+2x=5+x+8$
$3x+3=x+13$
$3(5)+3 \overset{?}{=} 5+13$
$15+3 \overset{?}{=} 5+13$
$18=18$
5 is a solution.

19. $\dfrac{3}{4}x=20$
$\dfrac{3}{4}(18) \overset{?}{=} 20$
$\dfrac{54}{4} \neq 20$
18 is not a solution.

21. $\dfrac{3}{5}x+5=11$
$\dfrac{3}{5}(10)+5 \overset{?}{=} 11$
$6+5 \overset{?}{=} 11$
$11=11$
10 is a solution.

23. $2x+1=9$ is a linear equation.

25. $2x-8$ is an expression.

27. $7x+2x+9-3$ is an expression.

29. $2x-8=3$ is a linear equation.

31. $x+9=11$
$\underline{-9 \quad -9}$
$x \quad = \quad 2$
Check:
$2+9 \overset{?}{=} 11$
$11=11$

33. $x-8 = \quad 3$
$\underline{+8 \quad +8}$
$x \quad = \quad 11$
Check:
$11-8 \overset{?}{=} 3$
$3=3$

35. $x-8=-10$
$\underline{+8 \quad +8}$
$x \quad = -2$
Check:
$-2-8 \overset{?}{=} -10$
$-10=-10$

37. $x+4=-3$
$\underline{-4 \quad -4}$
$x \quad =-7$
Check:
$-7+4 \overset{?}{=} -3$
$-3=-3$

39.
$$11 = x + 5$$
$$\underline{-5 \quad\;\; -5}$$
$$6 = x$$

Check:
$$11 \overset{?}{=} 6 + 5$$
$$11 = 11$$

41.
$$4x = 3x + 4$$
$$\underline{-3x \;\; -3x}$$
$$x = \quad\;\; 4$$

Check:
$$4(4) \overset{?}{=} 3(4) + 4$$
$$16 \overset{?}{=} 12 + 4$$
$$16 = 16$$

43.
$$11x = 10x - 10$$
$$\underline{-10x \;\; -10x}$$
$$x = -10$$

Check:
$$11(-10) \overset{?}{=} 10(-10) - 10$$
$$-110 \overset{?}{=} -100 - 10$$
$$-110 = -110$$

45.
$$6x + 3 = 5x$$
$$\underline{-5x \qquad -5x}$$
$$x + 3 = \;\; 0$$
$$\underline{\qquad -3 \;\; -3}$$
$$x \quad = -3$$

Check:
$$6(-3) + 3 \overset{?}{=} 5(-3)$$
$$-18 + 3 \overset{?}{=} -15$$
$$-15 = -15$$

47.
$$8x - 4 = \;\; 7x$$
$$\underline{-7x \qquad -7x}$$
$$x - 4 = \;\; 0$$
$$\underline{\quad +4 \;\; +4}$$
$$x \quad\; = \;\; 4$$

Check:
$$8(4) - 4 \overset{?}{=} 7(4)$$
$$32 - 4 \overset{?}{=} 28$$
$$28 = 28$$

49.
$$2x + 3 = \;\; x + 5$$
$$\underline{-x \qquad\; -x}$$
$$x + 3 = \;\; 5$$
$$\underline{\quad -3 \;\; -3}$$
$$x \qquad = \;\; 2$$

Check:
$$2(2) + 3 \overset{?}{=} 2 + 5$$
$$4 + 3 \overset{?}{=} 2 + 5$$
$$7 = 7$$

51.
$$4x - \frac{3}{5} = 3x + \frac{1}{10}$$
$$\underline{-3x \qquad\quad -3x}$$
$$x - \frac{3}{5} = \qquad \frac{1}{10}$$
$$\underline{\quad +\frac{3}{5} \qquad\quad +\frac{3}{5}}$$
$$x \qquad = \qquad \frac{7}{10}$$

Check:
$$4\left(\frac{7}{10}\right) - \frac{3}{5} \overset{?}{=} 3\left(\frac{7}{10}\right) + \frac{1}{10}$$
$$\frac{28}{10} - \frac{6}{10} \overset{?}{=} \frac{21}{10} + \frac{1}{10}$$
$$\frac{22}{10} = \frac{22}{10}$$

53.
$$\frac{7}{8}(x - 2) = \frac{3}{4} - \frac{1}{8}x$$
$$\frac{7}{8}x - \frac{14}{8} = \frac{3}{4} - \frac{1}{8}x$$
$$\underline{+\frac{1}{8}x \qquad\qquad\;\; +\frac{1}{8}x}$$
$$x - \frac{14}{8} = \frac{3}{4}$$
$$\underline{\quad +\frac{14}{8} \;\; +\frac{14}{8}}$$
$$x \qquad = \frac{20}{8}$$
$$x \qquad = \frac{5}{2}$$

Check:
$$\frac{7}{8}\left(\frac{5}{2} - 2\right) \overset{?}{=} \frac{3}{4} - \frac{1}{8}\left(\frac{5}{2}\right)$$
$$\frac{7}{8}\left(\frac{1}{2}\right) \overset{?}{=} \frac{3}{4} - \frac{5}{16}$$
$$\frac{7}{16} = \frac{7}{16}$$

55.
$$3x - 0.54 = 2(x - 0.15)$$
$$3x - 0.54 = 2x - 0.30$$

$$
\begin{array}{rcr}
-2x & & -2x \\
\hline
x - 0.54 = & & -0.30 \\
+0.54 & & +0.54 \\
\hline
x \quad = & & 0.24
\end{array}
$$

Check:
$$3(0.24) - 0.54 \overset{?}{=} 2(0.24 - 0.15)$$
$$0.72 - 0.54 \overset{?}{=} 2(0.09)$$
$$0.18 = 0.18$$

57.
$$6x + 3(x - 0.2789) = 4(2x + 0.3912)$$
$$6x + 3x - 0.8367 = 8x + 1.5648$$
$$9x - 0.8367 = 8x + 1.5648$$

$$
\begin{array}{rcr}
-8x & & -8x \\
\hline
x - 0.8367 = & & 1.5648 \\
+0.8367 & & +0.8367 \\
\hline
x \quad = & & 2.4015
\end{array}
$$

Check:
$$6(2.4015) + 3(2.4015 - 0.2789) \overset{?}{=} 4(2 \cdot 2.4015 + 0.3912)$$
$$14.409 + 3(2.1226) \overset{?}{=} 4(4.803 + 0.3912)$$
$$14.409 + 3(2.1226) \overset{?}{=} 4(5.1942)$$
$$20.7768 = 20.7768$$

59.
$$3x - 5 + 2x - 7 + x = 5x + 2$$
$$6x - 12 = 5x + 2$$

$$
\begin{array}{rcr}
-5x & & -5x \\
\hline
x - 12 = & & 2 \\
+12 & & +12 \\
\hline
x \quad = & & 14
\end{array}
$$

Check:
$$3(14) - 5 + 2(14) - 7 + 14 \overset{?}{=} 5(14) + 2$$
$$42 - 5 + 28 - 7 + 14 \overset{?}{=} 70 + 2$$
$$72 = 72$$

61.
$$5x - (0.345 - x) = 5x + 0.8713$$
$$5x - 0.345 + x = 5x + 0.8713$$
$$6x - 0.345 = 5x + 0.8713$$

$$
\begin{array}{rcr}
-5x & & -5x \\
\hline
x - 0.345 = & & 0.8713 \\
+0.345 & & +0.345 \\
\hline
x \quad = & & 1.2163
\end{array}
$$

Check:
$$5(1.2163) - (0.345 - 1.2163) \overset{?}{=} 5(1.2163) + 0.8713$$
$$6.0815 - (-0.8713) \overset{?}{=} 6.0815 + 0.8713$$
$$6.0815 + 0.8713 \overset{?}{=} 6.8015 + 0.8713$$
$$6.9528 = 6.9258$$

63.

$$3(7x+2) = 5(4x+1)+17$$
$$21x+6 = 20x+5+17$$
$$21x+6 = 20x+22$$

$$\underline{-20x \qquad\quad -20x}$$
$$x+6 = 22$$
$$\underline{-6 \qquad\qquad -6}$$
$$x = 16$$

Check:
$$3(7\cdot16+2)\overset{?}{=}5(4\cdot16+1)+17$$
$$3(112+2)\overset{?}{=}5(64+1)+17$$
$$3(114)\overset{?}{=}5(65)+17$$
$$342\overset{?}{=}325+17$$
$$342=342$$

65.

$$\frac{5}{4}x-1=\frac{1}{4}x+7$$
$$\underline{-\frac{1}{4}x \qquad\quad -\frac{1}{4}x}$$
$$x-1 = 7$$
$$\underline{+1 \qquad\quad +1}$$
$$x = 8$$

Check:
$$\frac{5}{4}(8)-1\overset{?}{=}\frac{1}{4}(8)+7$$
$$10-1\overset{?}{=}2+7$$
$$9=9$$

67.

$$\frac{9}{2}x-\frac{3}{4}=\frac{7}{2}x+\frac{5}{4}$$
$$\underline{-\frac{7}{2}x \qquad\qquad -\frac{7}{2}x}$$
$$x-\frac{3}{4}=\frac{5}{4}$$
$$\underline{+\frac{3}{4} \qquad\quad +\frac{3}{4}}$$
$$x = \frac{8}{4}$$
$$x = 2$$

Check:
$$\frac{9}{2}(2)-\frac{3}{4}\overset{?}{=}\frac{7}{2}(2)+\frac{5}{4}$$
$$9-\frac{3}{4}\overset{?}{=}7+\frac{5}{4}$$
$$\frac{36}{4}-\frac{3}{4}\overset{?}{=}\frac{28}{4}+\frac{5}{4}$$
$$\frac{33}{4}=\frac{33}{4}$$

69.

$$5x-7 = 4x-12$$
$$\underline{-4x \qquad\quad -4x}$$
$$x-7=-12$$

(d) is equivalent to the equation.

71.

$$7x+5=12x-10$$
$$\underline{+10 \qquad\quad +10}$$
$$7x+15=12x$$

(d) is equivalent to the equation.

73. It is false that isolating the variable one the right side of the equation will result in a negative solution.

75. Writing exercise

Exercises 2.2

1.

$$5x=20$$
$$\frac{5x}{5}=\frac{20}{5}$$
$$x=4$$
Check:
$$5(4)\overset{?}{=}20$$
$$20=20$$

3.

$$9x=54$$
$$\frac{9x}{9}=\frac{54}{9}$$
$$x=6$$
Check:
$$9(6)\overset{?}{=}54$$
$$54=54$$

5.

$$63=9x$$
$$\frac{63}{9}=\frac{9x}{9}$$
$$7=x$$
Check:
$$63\overset{?}{=}9(7)$$
$$63=63$$

7.

$$4x=-16$$
$$\frac{4x}{4}=\frac{-16}{4}$$
$$x=-4$$
Check:
$$4(-4)\overset{?}{=}-16$$
$$-16=-16$$

9. $-9x = 72$

$$\frac{-9x}{-9} = \frac{72}{-9}$$

$$x = -8$$

Check:

$-9(-8) \overset{?}{=} 72$

$72 = 72$

11. $6x = -54$

$$\frac{6x}{6} = \frac{-54}{6}$$

$$x = -9$$

Check:

$6(-9) \overset{?}{=} -54$

$-54 = -54$

13. $-4x = -12$

$$\frac{-4x}{-4} = \frac{-12}{-4}$$

$$x = 3$$

Check:

$-4(3) \overset{?}{=} -12$

$-12 = -12$

15. $-42 = 6x$

$$\frac{-42}{6} = \frac{6x}{6}$$

$$-7 = x$$

Check:

$-42 \overset{?}{=} 6(-7)$

$-42 = -42$

17. $-6x = -54$

$$\frac{-6x}{-6} = \frac{-54}{-6}$$

$$x = 9$$

Check:

$-6(9) \overset{?}{=} -54$

$-54 = -54$

19. $\dfrac{x}{2} = 4$

$$2\left(\frac{x}{2}\right) = 2 \cdot 4$$

$$x = 8$$

Check:

$\dfrac{8}{2} \overset{?}{=} 4$

$4 = 4$

21. $\dfrac{x}{5} = 3$

$$5\left(\frac{x}{5}\right) = 5 \cdot 3$$

$$x = 15$$

Check:

$\dfrac{15}{3} \overset{?}{=} 3$

$3 = 3$

23. $6 = \dfrac{x}{7}$

$$7 \cdot 6 = 7\left(\frac{x}{7}\right)$$

$$42 = x$$

Check:

$6 \overset{?}{=} \dfrac{42}{7}$

$6 = 6$

25. $\dfrac{x}{5} = -4$

$$5\left(\frac{x}{5}\right) = 5(-4)$$

$$x = -20$$

Check:

$\dfrac{-20}{5} \overset{?}{=} -4$

$-4 = -4$

27. $-\dfrac{x}{3} = 8$

$$-3\left(-\frac{x}{3}\right) = -3(8)$$

$$x = -24$$

Check:

$-\dfrac{-24}{3} \overset{?}{=} 8$

$8 = 8$

29. $\dfrac{2}{3}x = 0.9$

$$\frac{3}{2}\left(\frac{2}{3}x\right) = \frac{3}{2}(0.9)$$

$$x = 1.35$$

Check:

$\dfrac{2}{3}(1.35) \overset{?}{=} 0.9$

$0.9 = 0.9$

31.
$$\frac{3}{4}x = -15$$
$$\frac{4}{3}\left(\frac{3}{4}x\right) = \frac{4}{3}(-15)$$
$$x = -20$$
Check:
$$\frac{3}{4}(-20) \stackrel{?}{=} -15$$
$$-15 = -15$$

33.
$$-\frac{5}{6}x = -15$$
$$-\frac{6}{5}\left(-\frac{5}{6}x\right) = -\frac{6}{5}(-15)$$
$$x = 18$$
Check:
$$-\frac{5}{6}(18) \stackrel{?}{=} -15$$
$$-15 = -15$$

35. $16x - 9x = -16.1$
$$7x = -16.1$$
$$\frac{7x}{7} = \frac{-16.1}{7}$$
$$x = -2.3$$
Check:
$$16(-2.3) - 9(-2.3) \stackrel{?}{=} -16.1$$
$$-36.8 + 20.7 \stackrel{?}{=} -16.1$$
$$-16.1 = -16.1$$

37. $3.2x = 12.8$
$$\frac{3.2x}{3.2} = \frac{12.8}{3.2}$$
$$x = 4$$
Check:
$$3.2(4) \stackrel{?}{=} 12.8$$
$$12.8 = 12.8$$

39. $-4.5x = 3.51$
$$\frac{-4.5x}{-4.5} = \frac{3.51}{-4.5}$$
$$x = -0.78$$
Check:
$$-4.5(-0.78) \stackrel{?}{=} 3.51$$
$$3.51 = 3.51$$

41. $1.3x + 2.8x = 12.3$
$$4.1x = 12.3$$
$$\frac{4.1x}{4.1} = \frac{12.3}{4.1}$$
$$x = 3$$
Check:
$$1.3(3) + 2.8(3) \stackrel{?}{=} 12.3$$
$$3.9 + 8.4 \stackrel{?}{=} 12.3$$
$$12.3 = 12.3$$

43. The numbers are already in ascending order.
2, 3, 4, 5, 6
The median is the middle value, 4.
To find the mean, first find $\sum x$.
$$\sum x = 2 + 3 + 4 + 5 + 6 = 20$$
$$\bar{x} = \frac{\sum x}{n} = \frac{20}{5} = 4$$

45. The numbers are already in ascending order.
–3, –1, 2, 4, 6, 10
The median is $\frac{2+4}{2} = \frac{6}{2} = 3$.
To find the mean, first find $\sum x$.
$$\sum x = -3 + (-1) + 2 + 4 + 6 + 10 = 18$$
$$\bar{x} = \frac{\sum x}{n} = \frac{18}{6} = 3$$

47. The numbers are already in ascending order.
$-\frac{3}{2}, -1, 2, \frac{5}{2}, 3, 7$
The median is $\frac{2+\frac{5}{2}}{2} = \frac{\frac{9}{2}}{2} = \frac{9}{4}$.
To find the mean, first find $\sum x$.
$$\sum x = -\frac{3}{2} + (-1) + 2 + \frac{5}{2} + 3 + 7 = 12$$
$$\bar{x} = \frac{\sum x}{n} = \frac{12}{6} = 2$$

49. To find the median, first put the numbers in ascending order. 15, 16, 18, 21
The median is $\frac{16+18}{2} = 17$ oz.
To find the mean, first find $\sum x$.
$$\sum x = 15 + 16 + 18 + 21 = 70$$
$$\bar{x} = \frac{\sum x}{n} = \frac{70}{4} = 17.5 \text{ oz.}$$

Exercises 2.3

1.
$$2x + 1 = 9$$
$$\underline{-1 \quad -1}$$
$$2x \quad = 8$$
$$\frac{2x}{2} = \frac{8}{2}$$
$$x = 4$$
Check:
$$2(4) + 1 \stackrel{?}{=} 9$$
$$8 + 1 \stackrel{?}{=} 9$$
$$9 = 9$$

3.
$$3x - 2 = 7$$
$$\underline{+2 \quad +2}$$
$$3x \quad = 9$$
$$\frac{3x}{3} = \frac{9}{3}$$
$$x = 3$$
Check:
$$3(3) - 2 \stackrel{?}{=} 7$$
$$9 - 2 \stackrel{?}{=} 7$$
$$7 = 7$$

5.
$$4x + 7 = 35$$
$$\underline{-7 \quad -7}$$
$$4x \quad = 28$$
$$\frac{4x}{4} = \frac{28}{4}$$
$$x = 7$$
Check:
$$4(7) + 7 \stackrel{?}{=} 35$$
$$28 + 7 \stackrel{?}{=} 35$$
$$35 = 35$$

7.
$$2x + 9 = 5$$
$$\underline{-9 \quad -9}$$
$$2x \quad = -4$$
$$\frac{2x}{2} = \frac{-4}{2}$$
$$x = -2$$
Check:
$$2(-2) + 9 \stackrel{?}{=} 5$$
$$-4 + 9 \stackrel{?}{=} 5$$
$$5 = 5$$

9.
$$4 - 7x = 18$$
$$\underline{-4 -4}$$
$$-7x = 14$$
$$\frac{-7x}{-7} = \frac{14}{-7}$$
$$x = -2$$
Check:
$$4 - 7(-2) \stackrel{?}{=} 18$$
$$4 + 14 \stackrel{?}{=} 18$$
$$18 = 18$$

11.
$$3 - 4x = -9$$
$$\underline{-3 -3}$$
$$-4x = -12$$
$$\frac{-4x}{-4} = \frac{-12}{-4}$$
$$x = 3$$
Check:
$$3 - 4(3) \stackrel{?}{=} -9$$
$$3 - 12 \stackrel{?}{=} -9$$
$$-9 = -9$$

13.
$$\frac{x}{2} + 1 = 5$$
$$\underline{\phantom{\frac{x}{2}}-1 \quad -1}$$
$$\frac{x}{2} = 4$$
$$2\left(\frac{x}{2}\right) = (2)4$$
$$x = 8$$
Check:
$$\frac{8}{2} + 1 \stackrel{?}{=} 5$$
$$4 + 1 \stackrel{?}{=} 5$$
$$5 = 5$$

15.
$$\frac{x}{4} - 5 = 3$$
$$\underline{\phantom{\frac{x}{4}}+5 \quad +5}$$
$$\frac{x}{4} \quad = 8$$
$$4\left(\frac{x}{4}\right) = 4(8)$$
$$x = 32$$
Check:
$$\frac{32}{4} - 5 \stackrel{?}{=} 3$$
$$8 - 5 \stackrel{?}{=} 3$$
$$3 = 3$$

17. $\frac{2}{3}x + 5 = 17$

$$\underline{-5 \quad -5}$$

$$\frac{2}{3}x = 12$$

$$\frac{3}{2}\left(\frac{2}{3}x\right) = \frac{3}{2}(12)$$

$$x = 18$$

Check:

$$\frac{2}{3}(18) + 5 \stackrel{?}{=} 17$$

$$12 + 5 \stackrel{?}{=} 17$$

$$17 = 17$$

19. $\frac{4}{5}x - 3 = 13$

$$\underline{+3 \quad +3}$$

$$\frac{4}{5}x = 16$$

$$\frac{5}{4}\left(\frac{4}{5}x\right) = \frac{5}{4}(16)$$

$$x = 20$$

Check:

$$\frac{4}{5}(20) - 3 \stackrel{?}{=} 13$$

$$16 - 3 \stackrel{?}{=} 13$$

$$13 = 13$$

21. $5x = 2x + 9$

$$\underline{-2x \quad -2x}$$

$$3x = 9$$

$$\frac{3x}{3} = \frac{9}{3}$$

$$x = 3$$

Check:

$$5(3) \stackrel{?}{=} 2(3) + 9$$

$$15 \stackrel{?}{=} 6 + 9$$

$$15 = 15$$

23. $3x = 10 - 2x$

$$\underline{+2x +2x}$$

$$5x = 10$$

$$\frac{5x}{5} = \frac{10}{5}$$

$$x = 2$$

Check:

$$3(2) \stackrel{?}{=} 10 - 2(2)$$

$$6 \stackrel{?}{=} 10 - 4$$

$$6 = 6$$

25. $9x + 2 = 3x + 38$

$$\underline{-3x -3x}$$

$$6x + 2 = 38$$

$$\underline{-2 -2}$$

$$6x = 36$$

$$\frac{6x}{6} = \frac{36}{6}$$

$$x = 6$$

Check:

$$9(6) + 2 \stackrel{?}{=} 3(6) + 38$$

$$54 + 2 \stackrel{?}{=} 18 + 38$$

$$56 = 56$$

27. $4x - 8 = x - 14$

$$\underline{-x -x}$$

$$3x - 8 = -14$$

$$\underline{+8 +8}$$

$$3x = -6$$

$$\frac{3x}{3} = \frac{-6}{3}$$

$$x = -2$$

Check:

$$4(-2) - 8 \stackrel{?}{=} -2 - 14$$

$$-8 - 8 \stackrel{?}{=} -2 - 14$$

$$-16 = -16$$

29. $5x + 7 = 2x - 3$

$$\underline{-2x -2x}$$

$$3x + 7 = -3$$

$$\underline{-7 -7}$$

$$3x = -10$$

$$\frac{3x}{3} = \frac{-10}{3}$$

$$x = -\frac{10}{3}$$

Check:

$$5\left(-\frac{10}{3}\right) + 7 \stackrel{?}{=} 2\left(-\frac{10}{3}\right) - 3$$

$$-\frac{50}{3} + \frac{21}{3} \stackrel{?}{=} -\frac{20}{3} - \frac{9}{3}$$

$$-\frac{29}{3} = -\frac{29}{3}$$

31.
$$7x - 3 = 9x + 5$$
$$\underline{-7x \qquad -7x}$$
$$-3 = 2x + 5$$
$$\underline{-5 \qquad -5}$$
$$-8 = 2x$$
$$\frac{-8}{2} = \frac{2x}{2}$$
$$-4 = x$$

Check:
$$7(-4) - 3 \overset{?}{=} 9(-4) + 5$$
$$-28 - 3 \overset{?}{=} -36 + 5$$
$$-31 = -31$$

33.
$$5x + 4 = 7x - 8$$
$$\underline{-5x \qquad -5x}$$
$$4 = 2x - 8$$
$$\underline{+8 \qquad +8}$$
$$12 = 2x$$
$$\frac{12}{2} = \frac{2x}{2}$$
$$x = 6$$

Check:
$$5(6) + 4 \overset{?}{=} 7(6) - 8$$
$$30 + 4 \overset{?}{=} 42 - 8$$
$$34 = 34$$

35.
$$2x - 3 + 5x = 7 + 4x + 2$$
$$7x - 3 = 9 + 4x$$
$$\underline{-4x \qquad -4x}$$
$$3x - 3 = 9$$
$$\underline{+3 \quad +3}$$
$$3x = 12$$
$$\frac{3x}{3} = \frac{12}{3}$$
$$x = 4$$

Check:
$$2(4) - 3 + 5(4) \overset{?}{=} 7 + 4(4) + 2$$
$$8 - 3 + 20 \overset{?}{=} 7 + 16 + 2$$
$$25 = 25$$

37.
$$6x + 7 - 4x = 8 + 7x - 26$$
$$2x + 7 = 7x - 18$$
$$\underline{-2x \qquad -2x}$$
$$7 = 5x - 18$$
$$\underline{+18 \qquad +18}$$
$$25 = 5x$$
$$\frac{25}{5} = \frac{5x}{5}$$
$$x = 5$$

Check:
$$6(5) + 7 - 4(5) \overset{?}{=} 8 + 7(5) - 26$$
$$30 + 7 - 20 \overset{?}{=} 8 + 35 - 26$$
$$17 = 17$$

39.
$$9x - 2 + 7x + 13 = 10x - 13$$
$$16x + 11 = 10x - 13$$
$$\underline{-10x \qquad -10x}$$
$$6x + 11 = -13$$
$$\underline{-11 \quad -11}$$
$$6x = -24$$
$$\frac{6x}{6} = \frac{-24}{6}$$
$$x = -4$$

Check:
$$9(-4) - 2 + 7(-4) + 13 \overset{?}{=} 10(-4) - 13$$
$$-36 - 2 - 28 + 13 \overset{?}{=} -40 - 13$$
$$-53 = -53$$

41.
$$8x - 7 + 5x - 10 = 10x - 12$$
$$13x - 17 = 10x - 12$$
$$\underline{-10x \qquad = -10x}$$
$$3x - 17 = -12$$
$$\underline{+17 = \qquad +17}$$
$$3x = 5$$
$$\frac{3x}{3} = \frac{5}{3}$$
$$x = \frac{5}{3}$$

Check:
$$8\left(\frac{5}{3}\right) - 7 + 5\left(\frac{5}{3}\right) - 10 \overset{?}{=} 10\left(\frac{5}{3}\right) - 12$$
$$\frac{40}{3} - 7 + \frac{25}{3} - 10 \overset{?}{=} \frac{50}{3} - 12$$
$$\frac{14}{3} = \frac{14}{3}$$

Section 2.3

43. $7(2x-1)-5x = x+25$

$14x-7-5x = x+25$

$9x-7 = x+25$

$\underline{\quad -x \qquad -x \quad}$

$8x-7 = 25$

$\underline{\quad +7 \quad +7 \quad}$

$8x = 32$

$\dfrac{8x}{8} = \dfrac{32}{8}$

$x = 4$

Check:

$7(2\cdot 4-1)-5\cdot 4 \overset{?}{=} 4+25$

$49-20 \overset{?}{=} 29$

$29 = 29$

---

45. $3x+2(4x-3) = 6x-9$

$3x+8x-6 = 6x-9$

$11x-6 = 6x-9$

$\underline{\quad -6x \qquad -6x \quad}$

$5x-6 = -9$

$\underline{\quad +6 \qquad +6 \quad}$

$5x = -3$

$\dfrac{5x}{5} = \dfrac{-3}{5}$

$x = -\dfrac{3}{5}$

Check:

$3\left(-\dfrac{3}{5}\right)+2\left[4\left(-\dfrac{3}{5}\right)-3\right] \overset{?}{=} 6\left(-\dfrac{3}{5}\right)-9$

$-\dfrac{9}{5}+2\left(-\dfrac{27}{5}\right) \overset{?}{=} -\dfrac{18}{5}-\dfrac{45}{5}$

$-\dfrac{63}{5} = -\dfrac{63}{5}$

---

47. $\dfrac{8}{3}x-3 = \dfrac{2}{3}x+15$

$\underline{\qquad +3 \qquad\qquad +3 \quad}$

$\dfrac{8}{3}x = \dfrac{2}{3}x+18$

$\underline{\;\; -\dfrac{2}{3} \qquad -\dfrac{2}{3} \quad}$

$\dfrac{6}{3}x = 18$

$2x = 18$

$\dfrac{2x}{2} = \dfrac{18}{2}$

$x = 9$

Check:

$\dfrac{8}{3}(9)-3 \overset{?}{=} \dfrac{2}{3}(9)+15$

$24-3 \overset{?}{=} 6+15$

$21 = 21$

49. $\dfrac{2}{5}x-5 = \dfrac{12}{5}x+8$

$\underline{\; -\dfrac{2}{5}x \qquad\quad -\dfrac{2}{5}x \quad}$

$-5 = \dfrac{10}{5}x+8$

$\underline{\quad -8 \qquad\qquad -8 \quad}$

$-13 = \dfrac{10}{5}x$

$-13 = 2x$

$\dfrac{-13}{2} = \dfrac{2x}{2}$

$-\dfrac{13}{2} = x$

Check:

$\dfrac{2}{5}\left(-\dfrac{13}{2}\right)-5 \overset{?}{=} \dfrac{12}{5}\left(-\dfrac{13}{2}\right)+8$

$-\dfrac{13}{5}-\dfrac{25}{5} \overset{?}{=} -\dfrac{78}{5}+\dfrac{40}{5}$

$-\dfrac{38}{5} = -\dfrac{38}{5}$

---

51. $5.3x-7 = 2.3x+5$

$\underline{\; -2.3x \qquad -2.3x \quad}$

$3x-7 = 5$

$\underline{\quad +7 \qquad\quad +7 \quad}$

$3x = 12$

$\dfrac{3x}{3} = \dfrac{12}{3}$

$x = 4$

Check:

$5.3(4)-7 \overset{?}{=} 2.3(4)+5$

$21.2-7 \overset{?}{=} 9.2+5$

$14.2 = 14.2$

---

53. $4(x+5) = 4x+20$

$4x+20 = 4x+20$

$\underline{\; -4x \qquad -4x \quad}$

$20 = 20$

The original equation is an identity.

55. $5(x+1)-4x = x-5$

$5x+5-4x = x-5$

$x+5 = x-5$

$\underline{\quad -x \qquad -x \quad}$

$5 = -5$

The original equation has no solution.

57. $6x-4x+1 = 12+2x-11$

$2x+1 = 2x+1$

$\underline{-2x \qquad -2x \quad}$

$1 = 1$

The original equation is an identity.

59. $-4(x+2)-11 = 2(-2x-3)-13$

$-4x-8-11 = -4x-6-13$

$-4x-19 = -4x-19$

$\underline{+4x \qquad\quad +4x \quad}$

$-19 = -19$

The original equation is an identity.

61. Answers may vary.
Possible answer:

$x = 2$

$6x = 6 \cdot 2$

$6x = 12$

$6x+5 = 12+5$

$6x+5 = 17$

63. Writing exercise

65. $x+2x-2+x+2 = 24$

$4x = 24$

$\dfrac{4x}{4} = \dfrac{24}{4}$

$x = 6$

$2x-2 = 10$

$x+2 = 8$

The sides are 6 in., 8 in., and 10 in.

67. $3x-1+3x+2x-1+x+2 = 90$

$9x = 90$

$\dfrac{9x}{9} = \dfrac{90}{9}$

$x = 10$

$3x-1 = 29$

$3x = 30$

$2x-1 = 19$

$x+2 = 12$

The sides are 29 in., 30 in., 19 in., and 12 in.

Exercises 2.4

1. $p = 4s$

$\dfrac{p}{4} = \dfrac{4s}{4}$

$\dfrac{p}{4} = s$

3. $E = IR$

$\dfrac{E}{I} = \dfrac{IR}{I}$

$\dfrac{E}{I} = R$

5. $V = LWH$

$\dfrac{V}{LW} = \dfrac{LWH}{LW}$

$\dfrac{V}{LW} = H$

7. $A+B+C = 180$

$\underline{-A \quad\; -C \qquad -A-C \;}$

$B \quad = 180-A-C$

9. $ax+b = 0$

$\underline{\quad -b \quad -b \;}$

$ax = -b$

$\dfrac{ax}{a} = \dfrac{-b}{a}$

$x = -\dfrac{b}{a}$

11. $s = \frac{1}{2}gt^2$

$2s = 2\left(\frac{1}{2}gt^2\right)$

$2s = gt^2$

$\dfrac{2s}{t^2} = \dfrac{gt^2}{t^2}$

$\dfrac{2s}{t^2} = g$

13. $x+5y = 15$

$\underline{-x \qquad\quad -x \qquad}$

$5y = 15-x$

$\dfrac{5y}{5} = \dfrac{15-x}{5}$

$y = \dfrac{15-x}{5}$

or $y = -\dfrac{1}{5}x+3$

15.
$$P = 2L + 2W$$
$$\underline{ -2W \qquad -2W}$$
$$P - 2W = 2L$$
$$\frac{P - 2W}{2} = \frac{2L}{2}$$
$$L = \frac{P - 2W}{2}$$
$$\text{or } L = \frac{P}{2} - W$$

17.
$$V = \frac{KT}{P}$$
$$\frac{P}{K}(V) = \frac{P}{K}\left(\frac{KT}{P}\right)$$
$$\frac{PV}{K} = T$$

19.
$$x = \frac{a + b}{2}$$
$$2x = 2\left(\frac{a + b}{2}\right)$$
$$2x = a + b$$
$$\underline{ -a \qquad -a}$$
$$2x - a = b$$

21.
$$F = \frac{9}{5}C + 32$$
$$\underline{ -32 \qquad -32}$$
$$F - 32 = \frac{9}{5}C$$
$$\frac{5}{9}(F - 32) = \frac{5}{9}\left(\frac{9}{5}C\right)$$
$$C = \frac{5}{9}(F - 32)$$
$$\text{or } C = \frac{5(F - 32)}{9}$$

23.
$$S = 2\pi r^2 + 2\pi rh$$
$$\underline{ -2\pi r^2 \quad -2\pi r^2}$$
$$S - 2\pi r^2 = 2\pi rh$$
$$\frac{S - 2\pi r^2}{2\pi r} = \frac{2\pi rh}{2\pi r}$$
$$\frac{S - 2\pi r^2}{2\pi r} = h$$
$$\text{or } \frac{S}{2\pi r} - r = h$$

25. From Exercise 5:
$$H = \frac{V}{LW} = \frac{120}{(8)(5)} = 3$$
The height is 3 cm.

27. From Exercise 4:
$$r = \frac{I}{Pt} = \frac{450}{(3000)(3)} = 0.05$$
The interest rate is 5%.

29. From Exercise 21:
$$C = \frac{5(F - 32)}{9} = \frac{5(77 - 32)}{9} = \frac{5(45)}{9} = 25$$
The temperature would be given as 25°C.

31. $x + 3 = 7$

33. $3x - 7 = 2x$

35. $2(x + 5) = x + 18$

37. $2x + 3 = 7$

39. $4x - 7 = 41$

41. $\dfrac{2}{3}x + 5 = 21$

43. $3x = x + 12$

45. Let x be the number.
$$x + 7 = 33$$
$$x = 26$$
The number is 26.

47. Let x be the number.
$$x + (-15) = 7$$
$$x = 22$$
The number is 22.

49. Let x be the number of votes for the loser.
$$3260 - x = 1840$$
$$x = 1420$$
The losing candidate received 1420 votes.

51. Let x be the number.
$$2x + 7 = 33$$
$$2x = 26$$
$$x = 13$$
The number is 13.

53. Let x be the number.
$$5x - 12 = 78$$
$$5x = 90$$
$$x = 18$$

55. Let x be the 1st integer.
Then $x + 1$ is the next integer.
$$x + (x + 1) = 71$$
$$2x = 70$$
$$x = 35; \; x + 1 = 36$$
The integers are 35 and 36.

57. Let x be the 1st integer.
Then $x + 1$ is the 2nd integer, and
$x + 2$ is the 3rd integer.
$$x + (x + 1) + (x + 2) = 63$$
$$3x = 60$$
$$x = 20; \; x + 1 = 21; \; x + 2 = 22$$
The integers are 20, 21, and 22.

59. Let x be the 1st even integer.
Then $x + 2$ is the next even integer.
$$x + (x + 2) = 66$$
$$2x = 64$$
$$x = 32; \; x + 2 = 34$$
The integers are 32 and 34.

61. Let x be the 1st odd integer.
Then $x + 2$ is the next odd integer.
$$x + (x + 2) = 52$$
$$2x = 50$$
$$x = 25; \; x + 2 = 27$$
The integers are 25 and 27.

63. Let x be the 1st odd integer.
Then $x + 2$ is the 2nd odd integer, and
$x + 4$ is the 3rd odd integer.
$$x + (x + 2) + (x + 4) = 105$$
$$3x = 99$$
$$x = 33; \; x + 2 = 35; \; x + 4 = 37$$
The integers are 33, 35, and 37.

65. Let x be the 1st integer.
Then $x + 1$ is the 2nd integer,
$x + 2$ is the 3rd integer, and
$x + 3$ is the 4th integer.
$$x + (x + 1) + (x + 2) + x + 3 = 86$$
$$4x = 80$$
$$x = 20; \; x + 1 = 21; \; x + 2 = 22; \; x + 3 = 23$$
The integers are 20, 21, 22, and 23.

67. Let x be the 1st integer. Then $x + 1$ is the next integer.
$$4x = 3(x+1)+9$$
$$4x = 3x+12$$
$$x = 12;\ x+1 = 13$$
The integers are 12 and 13.

69. Let x be the number of votes for the loser and $x + 160$ be the number of votes for the winner.
$$x+(x+160) = 3260$$
$$2x = 3100$$
$$x = 1550;\ x+160 = 1710.$$
The winning candidate had 1710 votes and the loser had 1550 votes.

71. Let x be the cost of the dryer and $x + 70$ be the cost of the washer.
$$x+(x+70) = 650$$
$$2x = 580$$
$$x = 290;\ x+70 = 360.$$
The dryer costs $290 and the washer costs $360.

73. Let x be Yan's sister's age and $2x + 1$ be Yan's age.
$$x+(2x-1) = 14$$
$$3x = 15$$
$$x = 5;\ 2x-1 = 9$$
Yan is 9 years old.

75. Let x be Maritza's daughter's age and $4x - 3$ be Maritza's age.
$$x+(4x-3) = 37$$
$$5x = 40$$
$$x = 8;\ 4x-3 = 29$$
Maritza is 29 years old.

77. Let x be Jovita's airfare and $2x - 60$ be Jovita's food and lodging expenses.
$$x+(2x-60) = 2400$$
$$3x = 2460$$
$$x = 820$$
Her airfare was $820.

79. Let x be the number of students registered at 8 o'clock,
$2x$ be the number of students registered at 10 o'clock, and
$x + 7$ be the number of students registered at 12 o'clock.
$$x+2x+x+7 = 99$$
$$4x = 92$$
$$x = 23;\ 2x = 46;\ x+7 = 30$$
There are 23 students in the 8 o'clock section, 46 students in the 10 o'clock section, and 30 students in the 12 o'clock section.

81. Writing exercise

83. Writing exercise

Exercises 2.5

1. $3(x-5)=6$
$3x-15=6$
$3x=21$
$x=7$
Check: $3(7-5)=6$

3. $5(2x+3)=35$
$10x+15=35$
$10x=20$
$x=2$
Check: $5(2\cdot2+3)=35$

5. $7(5x+8)=-84$
$35x+56=-84$
$35x=-140$
$x=-4$
Check: $7(5(-4)+8)=-84$

7. $10-(x-2)=15$
$12-x=15$
$x=-3$
Check: $10-(-3-2)=15$

9. $5-(2x+1)=12$
$4-2x=12$
$-2x=8$
$x=-4$
Check: $5-(2(-4)+1)=12$

11. $7-(3x-5)=13$
$12-3x=13$
$-3x=1$
$x=-\dfrac{1}{3}$
Check: $7-\left(3\left(-\dfrac{1}{3}\right)-5\right)=13$

13. $5x=3(x-6)$
$5x=3x-18$
$2x=-18$
$x=-9$
Check: $5(-9)\overset{?}{=}3(-9-6)$
$-45=-45$

15. $7(2x-3)=20x$
$14x-21=20x$
$-21=6x$
$x=-\dfrac{7}{2}$
Check: $7\left(2\left(-\dfrac{7}{2}\right)-3\right)\overset{?}{=}20\left(-\dfrac{7}{2}\right)$
$-70=-70$

17. $6(6-x)=3x$
$36-6x=3x$
$36=9x$
$x=4$
Check: $6(6-4)\overset{?}{=}3(4)$
$12=12$

19. $2(2x-1)=3(x+1)$
$4x-2=3x+3$
$x=5$
Check: $2(2\cdot5-1)\overset{?}{=}3(5+1)$
$18=18$

21. $5(4x+2)=6(3x+4)$
$20x+10=18x+24$
$2x=14$
$x=7$
Check: $5(4\cdot7+2)\overset{?}{=}6(3\cdot7+4)$
$150=150$

23. $9(8x-1)=5(4x+6)$
$72x-9=20x+30$
$52x=39$
$x=\dfrac{3}{4}$
Check: $9\left(8\left(\dfrac{3}{4}\right)-1\right)\overset{?}{=}5\left(4\left(\dfrac{3}{4}\right)+6\right)$
$45=45$

25. $-4(2x-1)+3(3x+1)=9$
$-8x+4+9x+3=9$
$x=2$
Check: $-4(2\cdot2-1)+3(3\cdot2+1)=9$

27. $5(2x-1)-3(x-4)=4(x+4)$
$10x-5-3x+12=4x+16$
$7x+7=4x+16$
$x=3$
Check: $5(2\cdot3-1)-3(3-4))\overset{?}{=}4(3+4)$
$28=28$

29. $3(3-4x)+30=5x-2(6x-7)$
$9-12x+30=5x-12x+14$
$39-12x=-7x+14$
$-5x=-25$
$x=5$
Check: $3(3-4\cdot5)+30\overset{?}{=}5\cdot5-2(6\cdot5-7)$
$-21=-21$

31. $-2x+[3x-(-2x+5)]=-(15+2x)$
$-2x+(5x-5)=-15-2x$
$3x-5=-15-2x$
$5x=-10$
$x=-2$
Check: $-2(-2)+[3(-2)-(-2(-2)+5)]\overset{?}{=}-(15+2(-2))$
$-11=-11$

33. $3x^2-2(x^2+2)=x^2-4$
$3x^2-2x^2-4=x^2-4$
$x^2-4=x^2-4$
$0=0;$ all real numbers

35. Let x be the smaller number and $x+8$ be the larger number.
$x+2(x+8)=46$
$3x+16=46$
$3x=30$
$x=10;\ x+8=18$
The numbers are 10 and 18.

37. Let x be the larger number and $x-7$ be the smaller number.
$4(x-7)+2x=62$
$6x-28=62$
$6x=90$
$x=15;\ x-7=8$
The numbers are 8 and 15.

39. Let x be the 1st integer. Then $x+1$ is the next integer.
$2x+3(x+1)=28$
$5x+3=28$
$5x=25$
$x=5;\ x+1=6$
The numbers are 5 and 6.

41. Let x be width and $2x+1$ be length.
$P=2L+W$
$74=2(2x+1)+2x$
$74=6x+2$
$72=6x$
$x=12;\ 2x+1=25$
The width is 12 in. and the length is 25 in.

43. Let x be width of the garden and $3x + 4$ be length of the garden.

$P = 2L + 2W$
$56 = 2(3x + 4) + 2x$
$56 = 8x + 8$
$48 = 8x$
$x = 6; 3x + 4 = 22$

The width is 6 m and the length is 22 m.

45. Let x be length of the equal sides of the triangle and $x - 3$ be base of the triangle.

$P = A + B + C$
$36 = x + x + (x - 3)$
$36 = 3x - 3$
$39 = 3x$
$x = 13; x - 3 = 10$

The base is 10 cm and the length of the equal sides is 13 cm.

47. Let x be the number of main floor tickets and $500 - x$ be the number of balcony tickets.

$8x + 6(500 - x) = 3600$
$2x + 3000 = 3600$
$2x = 600$
$x = 300; 500 - x = 200$

There were 300 main floor tickets and
200 balcony tickets sold.

49. Let x be the amount of $0.33 denomination stamps and $80 - x$ be amount of $0.25 denomination stamps.

$0.33x + 0.25(80 - x) = 24$
$0.33x + 20 - 0.25x = 24$
$20 + 0.08 = 24$
$0.08x = 4$
$x = 50; 80 - x = 30$

She bought 50 33¢ stamps and 30 25¢ stamps.

51. Let x be number of tickets for the sleeping rooms,
$x + 20$ be number of tickets for the berths, and
$3x$ be number of tickets for the coach seats.

$120x + 80(x + 20) + 50(3x) = 8600$
$350x + 1600 = 8600$
$350x = 7000$
$x = 20; x + 20 = 40; 3x = 60$

There were 20 sleeping room, 40 berth, and 60 coach tickets sold.

53.

	Distance	Rate	Time
Going	$3r$	r	3
Returning	$4(r-10)$	$r-10$	4

distance going = distance returning

$$3r = 4(r-10)$$
$$3r = 4r - 40$$
$$r = 40;\ r - 10 = 30$$

His speed was 40 mph going and 30 mph returning.

55.

	Distance	Rate	Time
First Car	$50t$	50	t
Second Car	$40(t-1)$	40	$t-1$

distance between cars = 320

$$50t + 40(t-1) = 320$$
$$90t - 40 = 320$$
$$90t = 360$$
$$t = 4 \text{ hrs past 2 P. M.}$$

They will be 320 miles apart at 6 P.M.

57.

	Distance	Rate	Time
Catherine	$45t$	45	t
Max	$54(t-1)$	54	$t-1$

Max's distance = Catherine's distance

$$45t = 54(t-1)$$
$$45t = 54t - 54$$
$$54 = 9t$$
$$t = 6 \text{ hrs past 8 A.M.}$$

Max will catch up with Catherine at 2 P.M.

59.

	Distance	Rate	Time
Mika	$45t$	45	t
Hiroka	$50(t-1)$	50	$t-1$

Mika's distance + Hiroko's distance = 425

$$45t + 50(t-1) = 425$$
$$95t - 50 = 425$$
$$95t = 475$$
$$t = 5 \text{ hrs past 10 A.M.}$$

They will meet at 3 P.M.

61. Let x be number of Douglas fir trees and $500 - x$ be number of hemlock trees.

$$250x + 300(500 - x) = 132,000$$
$$250x + 150,000 - 300x = 132,000$$
$$-50x = -18,000$$
$$x = 360; \ 500 - x = 140$$

They bought 360 Douglas fir and 140 hemlock trees.

63. Writing exercise

65. Writing exercise

Exercises 2.6

1. $0.23 = 23\%$

3. $2.5 = 2.50 = 250\%$

5. $\dfrac{3}{8} = 0.375 = 37.5\%$

7. The rate in the statement "23% of 400 is 92" is 23%.

9. The amount in the statement "200 is 40% of 500" is 200.

11. The base in the statement "16% of 350 is 56" is 350.

13. The rate is 5%.
The base is $40,000.
The amount is "what commission".

15. The rate is "what percent".
The base is 30.
The amount is 5.

17. The rate is 5%.
The base is "selling price".
The amount is 3.30.

19. The rate is 6%
The base is 9000.
The amount is "students".

21. $\dfrac{A}{3400} = \dfrac{12}{100}$

$$100A = 12 \cdot 3400$$
$$A = \dfrac{40,800}{100}$$
$$A = 408$$

The interest is $408.

23. $\dfrac{A}{550} = \dfrac{26}{100}$

$$100A = 26 \cdot 550$$
$$A = \dfrac{14,300}{100}$$
$$A = 143$$

$143 is withheld per week.

25. $\dfrac{140}{2800} = \dfrac{R}{100}$

$$2800R = 140 \cdot 100$$
$$R = \dfrac{14,000}{2800}$$
$$R = 5$$

The commission rate is 5%.

27. $\dfrac{18}{1200} = \dfrac{R}{100}$

$$1200R = 18 \cdot 100$$
$$R = \dfrac{1800}{1200}$$
$$R = 1.5$$

The interest rate is 1.5%.

29. $\dfrac{20}{B} = \dfrac{80}{100}$

$$80B = 20 \cdot 100$$
$$B = \dfrac{2000}{80}$$
$$B = 25$$

There were 25 questions.

31. $\dfrac{525}{B} = \dfrac{10.5}{100}$

$$10.5B = 525 \cdot 100$$
$$B = \dfrac{52,500}{10.5}$$
$$B = 5000$$

The loan was for $5000.

33. $\dfrac{A}{260} = \dfrac{6.4}{100}$

$100A = 260(6.4)$

$A = \dfrac{1664}{100}$

$A = 16.64$

The tax would be \$16.64.

35. $\dfrac{A}{125,000} = \dfrac{6.5}{100}$

$100A = 6.5(125,000)$

$A = \dfrac{812,500}{100}$

$A = 8125$

The salesperson will make \$8125.

37. $\dfrac{102}{1200} = \dfrac{R}{100}$

$1200R = 102 \cdot 100$

$R = \dfrac{10,200}{1200}$

$R = 8.5$

The unemployment rate is 8.5%

39. $\dfrac{45}{60} = \dfrac{R}{100}$

$60R = 45 \cdot 100$

$R = \dfrac{4500}{60}$

$R = 75$

75% completed the course,
so 100% − 75% = 25% dropped out.

41. $\dfrac{780}{B} = \dfrac{65}{100}$

$65B = 780 \cdot 100$

$B = \dfrac{78,000}{65}$

$B = 1200$

1200 people responded to the survey.

43. $\dfrac{209}{B} = \dfrac{22}{100}$

$22B = 209 \cdot 100$

$B = \dfrac{20,900}{22}$

$B = 950$

His salary is \$950.

45. $\dfrac{A}{600} = \dfrac{22}{100}$

$100A = 22 \cdot 600$

$A = \dfrac{13,200}{100}$

$A = 132$

$600 + 132 = 732$

The selling price is \$732.

47. $\dfrac{A}{125,000} = \dfrac{25}{100}$

$100A = 25(125,000)$

$A = \dfrac{3,125,000}{100}$

$A = 31,250$

$125,000 + 31,250 = 156,250$

The lot's value is \$156,250.

49. increase = 1064 − 950 = 114

$\dfrac{114}{950} = \dfrac{R}{100}$

$950R = 114 \cdot 100$

$R = \dfrac{11,400}{950}$

$R = 12$

The rate of increase is 12%.

51. discount = 450 − 382.5 = 67.5

$\dfrac{67.5}{450} = \dfrac{R}{100}$

$450R = 67.5(100)$

$R = \dfrac{6750}{450}$

$R = 15$

The discount rate is 15%.

53. $\dfrac{4830}{B} = \dfrac{14}{100}$

$14 = 4830(100)$

$B = \dfrac{483,000}{14}$

$B = 34,500$

The price was \$34,500 before the increase.

55. $\dfrac{66}{B} = \dfrac{5.5}{100}$

$5.5B = 66(100)$

$B = \dfrac{6600}{5.5}$

$B = 1200$

The company had 1200 employees in
July 1998.

57. 25% off means they are 75% of the original price.

$$\frac{48.75}{B} = \frac{75}{100}$$

$$75B = 48.75(100)$$

$$B = \frac{4875}{75}$$

$$B = 65$$

The original price was $65.

59. increase = 71,388 − 40,592 = 30,796

$$\frac{30,796}{40,592} = \frac{R}{100}$$

$$40,592R = 30,796(100)$$

$$R = 76$$

The rate of increase is 76%.

61. difference = 40,592 − 35,211 = 5381

$$\frac{5381}{35,211} = \frac{R}{100}$$

$$35,211R = 5381(100)$$

$$R \approx 15$$

Exports exceeded imports by about 15%.

63. $0.06 \times 4000 = 240$

  $4000 + 240 = \$4240$ after 1 year

  $0.06 \times 4240 = 254.4$

  $4240 + 254.4 = \$4494.40$ after 2 years.

65. $0.05 \times 4000 = 200$

  $4000 + 200 = \$4200$ after 1 year

  $0.05 \times 4200 = 210$

  $4200 + 210 = \$4410$ after 2 years

  $0.05 \times 4410 = 220.50$

  $4410 + 220.50 = \$4630.50$ after 3 years.

67. $\dfrac{145.0}{194.5} = \dfrac{R}{100}$

  $194.5R = 145.0(100)$

  $R \approx 74.6$

About 74.6% of the vehicles registered were passenger cars.

69. $\dfrac{10.85}{B} = \dfrac{63}{100}$

  $63B = 10.85(100)$

  $B \approx 17.22$

About 17.22 million bbl of petroleum are consumed by the United States.

71. $\dfrac{15,000}{40,000} = \dfrac{R}{100}$

  $40,000R = 15,000(100)$

  $R = 37.5$

37.5% of the goal has been achieved so far.

73. $\overline{AD} = 12,\ \overline{AB} = 16$

  $\dfrac{12}{16} = \dfrac{R}{100}$

  $16R = 12(100)$

  $R = 75$

Length AD is 75% of length AB.

75. $\overline{AE} = 6,\ \overline{AD} = 12$

  $\dfrac{6}{12} = \dfrac{R}{100}$

  $12R = 6(100)$

  $R = 50$

Length AE is 50% of length AD.

Exercises 2.7

1. $5 < 10$

3. $7 > -2$

5. $0 < 4$

7. $-2 > -5$

9. $x < 3$: x is less than 3

11. $x \geq -4$: x is greater than or equal to −4

13. $-5 \leq x$: −5 is less than or equal to x

15. $x > 2$

17. $x < 9$

19. $x > 1$

21. $x < 8$

23. $x > -5$

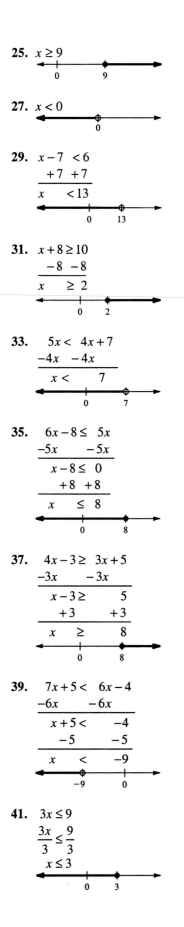

25. $x \geq 9$

27. $x < 0$

29. $x - 7 < 6$
$$\frac{+7 \quad +7}{x \quad < 13}$$

31. $x + 8 \geq 10$
$$\frac{-8 \quad -8}{x \quad \geq 2}$$

33. $5x < 4x + 7$
$$\frac{-4x \quad -4x}{x < \quad 7}$$

35. $6x - 8 \leq 5x$
$$\frac{-5x \qquad -5x}{x - 8 \leq 0}$$
$$\frac{+8 \quad +8}{x \quad \leq 8}$$

37. $4x - 3 \geq 3x + 5$
$$\frac{-3x \qquad -3x}{x - 3 \geq \qquad 5}$$
$$\frac{+3 \qquad +3}{x \quad \geq \qquad 8}$$

39. $7x + 5 < 6x - 4$
$$\frac{-6x \qquad -6x}{x + 5 < \qquad -4}$$
$$\frac{-5 \qquad -5}{x \quad < \qquad -9}$$

41. $3x \leq 9$
$$\frac{3x}{3} \leq \frac{9}{3}$$
$$x \leq 3$$

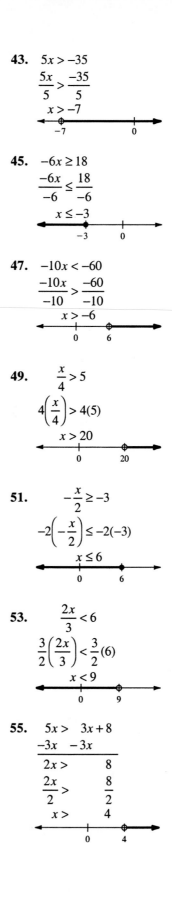

43. $5x > -35$
$$\frac{5x}{5} > \frac{-35}{5}$$
$$x > -7$$

45. $-6x \geq 18$
$$\frac{-6x}{-6} \leq \frac{18}{-6}$$
$$x \leq -3$$

47. $-10x < -60$
$$\frac{-10x}{-10} > \frac{-60}{-10}$$
$$x > -6$$

49. $\dfrac{x}{4} > 5$
$$4\left(\frac{x}{4}\right) > 4(5)$$
$$x > 20$$

51. $-\dfrac{x}{2} \geq -3$
$$-2\left(-\frac{x}{2}\right) \leq -2(-3)$$
$$x \leq 6$$

53. $\dfrac{2x}{3} < 6$
$$\frac{3}{2}\left(\frac{2x}{3}\right) < \frac{3}{2}(6)$$
$$x < 9$$

55. $5x > 3x + 8$
$$\frac{-3x \quad -3x}{2x > \qquad 8}$$
$$\frac{2x}{2} > \frac{8}{2}$$
$$x > \qquad 4$$

57.
$$5x - 2 > 3x$$
$$\underline{-3x \qquad -3x}$$
$$2x - 2 > 0$$
$$\underline{+2 \quad +2}$$
$$2x \qquad > 2$$
$$\frac{2x}{2} > \frac{2}{2}$$
$$x \qquad > 1$$

59.
$$3 - 2x > 5$$
$$\underline{-3 \qquad -3}$$
$$-2x > 2$$
$$\frac{-2x}{-2} < \frac{2}{-2}$$
$$x < -1$$

61.
$$2x \geq 5x + 18$$
$$\underline{-5x \quad -5x}$$
$$-3x \geq 18$$
$$\frac{-3x}{-3} \leq \frac{18}{-3}$$
$$x \leq -6$$

63.
$$5x - 3 \leq 3x + 15$$
$$\underline{-3x \qquad -3x}$$
$$2x - 3 \leq 15$$
$$\underline{+3 \qquad +3}$$
$$2x \leq 18$$
$$\frac{2x}{2} \leq \frac{18}{2}$$
$$x \leq 9$$

65.
$$9x + 7 > 2x - 28$$
$$\underline{-2x \qquad -2x}$$
$$7x + 7 > -28$$
$$\underline{-7 \qquad -7}$$
$$7x > -35$$
$$\frac{7x}{7} > \frac{-35}{7}$$
$$x > -5$$

67.
$$7x - 5 < 3x + 2$$
$$\underline{-3x \qquad -3x}$$
$$4x - 5 < 2$$
$$\underline{+5 \qquad +5}$$
$$4x < 7$$
$$\frac{4x}{4} < \frac{7}{4}$$
$$x < \frac{7}{4}$$

69.
$$5x + 7 > 8x - 17$$
$$\underline{-8x \qquad -8x}$$
$$-3x + 7 > -17$$
$$\underline{-7 \qquad -7}$$
$$-3x > -24$$
$$\frac{-3x}{-3} < \frac{-24}{-3}$$
$$x < 8$$

71.
$$3x - 2 \leq 5x + 3$$
$$\underline{-5x \qquad -5x}$$
$$-2x - 2 \leq 3$$
$$\underline{+2 \qquad +2}$$
$$-2x \leq 5$$
$$\frac{-2x}{-2} \geq \frac{5}{-2}$$
$$x \geq -\frac{5}{2}$$

73.
$$4(x + 7) \leq 2x + 31$$
$$4x + 28 \leq 2x + 31$$
$$\underline{-2x \qquad -2x}$$
$$2x + 28 \leq 31$$
$$\underline{-28 \qquad -28}$$
$$2x \leq 3$$
$$\frac{2x}{2} \leq \frac{3}{2}$$
$$x \leq \frac{3}{2}$$

43

75.
$$2(x-7) > 5x-12$$
$$2x-14 > 5x-12$$

$$\underline{-5x \qquad -5x}$$
$$\underline{-3x-14 > \qquad -12}$$
$$\underline{+14 \qquad +14}$$
$$\underline{-3x \qquad > \qquad 2}$$
$$\frac{-3x}{-3} < \frac{2}{-3}$$
$$x < -\frac{2}{3}$$

77. $x+5 > 3$

79. $2x-4 \le 7$

81. $4x-15 > x$

83. x is nonnegative: $x \ge 0$; choice a

85. x is no more than 5: $x \ge 5$; choice c

87. x is at least 5: $x \ge 5$; choice b

89. P = panda population
$P < 1000$

91. Let x be grade on fourth test.
$$\frac{72+81+79+x}{4} \ge 80$$
$$232+x \ge 320$$
$$x \ge 88$$
Liza must earn a grade of at least 88 on the last test.

93. Let x be amount of sales needed.
$$0.05x > 500$$
$$x > 10,000$$
She needs to sell more than \$10,000 to make the 5% offer a better deal.

95. Let x be width.
$$2(105+x) \le 250$$
$$210+2x \le 250$$
$$2x \le 40$$
$$x \le 20$$
The width is to be no greater than 20 cm.

97. Writing exercise

Summary Exercises for Chapter 2

1.
$$7x+2 = 16$$
$$7(2)+2 \stackrel{?}{=} 16$$
$$14+2 \stackrel{?}{=} 16$$
$$16 = 16$$
2 is a solution

3.
$$7x-2 = 2x+8$$
$$7(2)-2 \stackrel{?}{=} 2(2)+8$$
$$14-2 \stackrel{?}{=} 4+8$$
$$12 = 12$$
2 is a solution.

5.
$$x+5+3x = 2+x+23$$
$$6+5+3(6) \stackrel{?}{=} 2+6+23$$
$$6+5+18 \stackrel{?}{=} 2+6+23$$
$$29 \ne 31$$
6 is not a solution.

7.
$$x+5 = 7$$
$$\underline{-5 \quad -5}$$
$$x \quad = 2$$
Check:
$$2+5 \stackrel{?}{=} 7$$
$$7 = 7$$

9.
$$5x = 4x-5$$
$$\underline{-4x \quad -4x}$$
$$x = -5$$
Check:
$$5(-5) \stackrel{?}{=} 4(-5)-5$$
$$-25 \stackrel{?}{=} -20-5$$
$$-25 = 25$$

11.
$$5x-3 = 4x+2$$
$$\underline{-4x \qquad -4x}$$
$$x-3 = 2$$
$$\underline{+3 \qquad +3}$$
$$x \quad = 5$$
Check:
$$5(5)-3 \stackrel{?}{=} 4(5)+2$$
$$25-3 \stackrel{?}{=} 20+2$$
$$22 = 22$$

13.
$$7x - 5 = 6x - 4$$
$$\underline{-6x \qquad -6x}$$
$$x - 5 = \quad -4$$
$$\underline{+5 \qquad +5}$$
$$x = \quad 1$$

Check:
$$7 \cdot 1 - 5 \overset{?}{=} 6 \cdot 1 - 4$$
$$7 - 5 \overset{?}{=} 6 - 4$$
$$2 = 2$$

15.
$$4(2x + 3) = 7x + 5$$
$$8x + 12 = 7x + 5$$
$$\underline{-7x \qquad -7x}$$
$$x + 12 = \quad 5$$
$$\underline{-12 \qquad -12}$$
$$x = \quad -7$$

Check:
$$4[2(-7) + 3] \overset{?}{=} 7(-7) + 5$$
$$4(-14 + 3) \overset{?}{=} -49 + 5$$
$$-44 = -44$$

17.
$$5x = 35$$
$$\frac{5x}{5} = \frac{35}{5}$$
$$x = 7$$
Check:
$$5 \cdot 7 \overset{?}{=} 35$$
$$35 = 35$$

19.
$$-6x = 24$$
$$\frac{-6x}{-6} = \frac{24}{-6}$$
$$x = -4$$
Check:
$$-6(-4) \overset{?}{=} 24$$
$$24 = 24$$

21.
$$\frac{x}{4} = 8$$
$$4\left(\frac{x}{4}\right) = 4(8)$$
$$x = 32$$
Check:
$$\frac{32}{4} \overset{?}{=} 8$$
$$8 = 8$$

23.
$$\frac{2}{3}x = 18$$
$$\frac{3}{2}\left(\frac{2}{3}x\right) = \frac{3}{2}(18)$$
$$x = 27$$
Check:
$$\frac{2}{3}(27) \overset{?}{=} 18$$
$$18 = 18$$

25.
$$5x - 3 = 12$$
$$\underline{+3 \quad +3}$$
$$5x = 15$$
$$\frac{5x}{5} = \frac{15}{5}$$
$$x = 3$$
Check:
$$5(3) - 3 \overset{?}{=} 12$$
$$15 - 3 \overset{?}{=} 12$$
$$12 = 12$$

27.
$$7x + 8 = 3x$$
$$\underline{-7x \qquad -7x}$$
$$8 = -4x$$
$$\frac{8}{-4} = \frac{-4x}{-4}$$
$$-2 = x$$
Check:
$$7(-2) + 8 \overset{?}{=} 3(-2)$$
$$-14 + 8 \overset{?}{=} -6$$
$$-6 = -6$$

29.
$$3x - 7 = x$$
$$\underline{-3x \qquad -3x}$$
$$-7 = -2x$$
$$\frac{-7}{-2} = \frac{-2x}{-2}$$
$$\frac{7}{2} = x$$
Check:
$$3\left(\frac{7}{2}\right) - 7 \overset{?}{=} \frac{7}{2}$$
$$\frac{7}{2} = \frac{7}{2}$$

31.

$$\frac{x}{3} - 5 = 1$$
$$\underline{+5 \quad +5}$$
$$\frac{x}{3} = 6$$
$$3\left(\frac{x}{3}\right) = 3(6)$$
$$x = 18$$

Check:
$$\frac{18}{3} - 5 \stackrel{?}{=} 1$$
$$6 - 5 \stackrel{?}{=} 1$$
$$1 = 1$$

33.

$$6x - 5 = 3x + 13$$
$$\underline{-3x \qquad -3x}$$
$$3x - 5 = \qquad 13$$
$$\underline{+5 \qquad +5}$$
$$3x = \qquad 18$$
$$\frac{3x}{3} = \frac{18}{3}$$
$$x = 6$$

Check:
$$6 \cdot 6 - 5 \stackrel{?}{=} 3 \cdot 6 + 13$$
$$36 - 5 \stackrel{?}{=} 18 + 13$$
$$31 = 31$$

35.

$$7x + 4 = 2x + 6$$
$$\underline{-2x \qquad -2x}$$
$$5x + 4 = \qquad 6$$
$$\underline{-4 \qquad -4}$$
$$5x = \qquad 2$$
$$\frac{5x}{5} = \frac{2}{5}$$
$$x = \frac{2}{5}$$

Check:
$$7\left(\frac{2}{5}\right) + 4 \stackrel{?}{=} 2\left(\frac{2}{5}\right) + 6$$
$$\frac{14}{5} + \frac{20}{5} \stackrel{?}{=} \frac{4}{5} + \frac{30}{5}$$
$$\frac{34}{5} = \frac{34}{5}$$

37.

$$2x + 7 = 4x - 5$$
$$\underline{-2x \qquad -2x}$$
$$7 = 2x - 5$$
$$\underline{+5 \qquad +5}$$
$$12 = 2x$$
$$\frac{12}{2} = \frac{2x}{2}$$
$$6 = x$$

Check:
$$2 \cdot 6 + 7 \stackrel{?}{=} 4 \cdot 6 - 5$$
$$12 + 7 \stackrel{?}{=} 24 - 5$$
$$19 = 19$$

39.

$$\frac{10}{3}x - 5 = \frac{4}{3}x + 7$$
$$\underline{-\frac{4}{3} \qquad -\frac{4}{3}x}$$
$$\frac{6}{3}x - 5 = \qquad 7$$
$$\underline{+5 \qquad +5}$$
$$\frac{6}{3}x = \qquad 12$$
$$2x = 12$$
$$\frac{2x}{2} = \frac{12}{2}$$
$$x = 6$$

Check:
$$\frac{10}{3}(6) - 5 \stackrel{?}{=} \frac{4}{3}(6) + 7$$
$$20 - 5 \stackrel{?}{=} 8 + 7$$
$$15 = 15$$

41.

$$3.7x + 8 = 1.7x + 16$$
$$\underline{-1.7x \qquad -1.7x}$$
$$2x + 8 = \qquad 16$$
$$\underline{-8 \qquad -8}$$
$$2x = \qquad 8$$
$$\frac{2x}{2} = \frac{8}{2}$$
$$x = 4$$

Check:
$$3.7(4) + 8 \stackrel{?}{=} 1.7(4) + 16$$
$$14.8 + 8 \stackrel{?}{=} 6.8 + 16$$
$$22.8 = 22.8$$

43.
$$3x - 2 + 5x = 7 + 2x + 21$$
$$8x - 2 = 2x + 28$$
$$\underline{-2x \qquad\qquad -2x}$$
$$\overline{6x - 2 \quad = \quad\qquad 28}$$
$$\underline{+2 \qquad\qquad +2}$$
$$\overline{6x \qquad = \qquad\qquad 30}$$
$$\frac{6x}{6} = \frac{30}{6}$$
$$x = 5$$

Check:
$$3 \cdot 5 - 2 + 5 \cdot 5 \overset{?}{=} 7 + 2 \cdot 5 + 21$$
$$15 - 2 + 25 \overset{?}{=} 7 + 10 + 21$$
$$38 = 38$$

45.
$$5(3x - 1) - 6x = 3x - 2$$
$$15x - 5 - 6x = 3x - 2$$
$$9x - 5 = 3x - 2$$
$$\underline{-3x \qquad -3x}$$
$$\overline{6x - 5 = \qquad -2}$$
$$\underline{+5 \qquad\quad +5}$$
$$\overline{6x \quad = \qquad 3}$$
$$\frac{6x}{6} = \frac{3}{6}$$
$$x = \frac{1}{2}$$

Check:
$$5\left(3 \cdot \frac{1}{2} - 1\right) - 6 \cdot \frac{1}{2} \overset{?}{=} 3 \cdot \frac{1}{2} - 2$$
$$5\left(\frac{1}{2}\right) - 3 \overset{?}{=} \frac{3}{2} - 2$$
$$-\frac{1}{2} = -\frac{1}{2}$$

47.
$$P = 2L + 2W$$
$$\underline{-2W \qquad -2W}$$
$$\overline{P - 2W = 2L}$$
$$\frac{P - 2W}{2} = L$$
$$\text{or } \frac{P}{2} - W = L$$

49.
$$A = \frac{1}{2}bh$$
$$\frac{2}{b}(A) = \frac{2}{b}\left(\frac{1}{2}bh\right)$$
$$\frac{2A}{b} = h$$

51.
$$m = \frac{n - p}{q}$$
$$m \cdot q = q \cdot \frac{n - p}{q}$$
$$mq = n - p$$
$$mq + p = n$$

53. Let x be the number.
$$5x - 8 = 32$$
$$5x = 40$$
$$x = 8$$

55. Let x be first odd integer,
$x + 2$ be second odd integer, and
$x + 4$ be third odd integer.
$$x + (x + 2) + (x + 4) = 57$$
$$3x + 6 = 57$$
$$3x = 51$$
$$x = 17; \; x + 2 = 19; \; x + 4 = 21$$
The integers are 17, 19, and 21.

57. Let x be Susan's age,
$x + 2$ be Larry's age, and
$2x$ be Nathan's age.
$$x + (x + 2) + 2x = 30$$
$$4x + 2 = 30$$
$$4x = 28$$
$$x = 7; \; x + 2 = 9; \; 2x = 14$$
Susan is 7 years, Larry is 9 years, and Nathan is 14 years.

59.
$$\frac{77}{350} = \frac{R}{100}$$
$$350R = 77(100)$$
$$R = \frac{7700}{350}$$
$$R = 22$$
The discount rate is 22%.

61.
$$\frac{819}{B} = \frac{4.5}{100}$$
$$4.5B = 819(100)$$
$$B = \frac{81,900}{4.5}$$
$$B = 18,200$$
The car was $18,200 before the increase.

63. $\dfrac{168}{B} = \dfrac{6}{100}$

$6B = 168(100)$

$B = \dfrac{16,800}{6}$

$B = 2800$

Tom's monthly salary is $2800.

65. increase $= 76,680 - 72,000 = 4680$

$\dfrac{4680}{72,2000} = \dfrac{R}{100}$

$72,000R = 4680(100)$

$R = \dfrac{468,000}{72,000}$

$R = 6.5$

The rate of increase was 6.5%.

67. $\dfrac{126}{B} = \dfrac{4}{100}$

$4B = 126(100)$

$B = \dfrac{12,600}{4}$

$B = 3150$

$3150 + 126 = 3276$

Her monthly salary was $3150 before and $3276 after.

69. $\dfrac{150}{B} = \dfrac{30}{100}$

$30B = 150(100)$

$B = \dfrac{15,000}{30}$

$B = 500$

It will take 500 seconds to check all the files.

71. $80.15 is 70% of the original cost.

$\dfrac{80.15}{B} = \dfrac{70}{100}$

$70B = 80.15(100)$

$B = \dfrac{8015}{70}$

$B = 114.50$

The original price was $114.50.

73. $x + 3 > -2$

$\quad\ \underline{-3 \quad -3}$

$x \qquad > -5$

75. $4x \ge -12$

$\dfrac{4x}{4} \ge \dfrac{-12}{4}$

$x \ge -3$

77. $-\dfrac{x}{5} \ge 3$

$-5\left(-\dfrac{x}{5}\right) \le -5(3)$

$x \le -15$

79. $2x + 3 \ge 9$

$\quad\ \underline{-3 \quad -3}$

$2x \quad \ge \ 6$

$\dfrac{2x}{2} \ge \dfrac{6}{2}$

$x \ge 3$

81. $5x - 2 \le 4x + 5$

$\underline{-4x \qquad\ -4x}$

$x - 2 \le \qquad 5$

$\underline{\ +2 \qquad\ +2}$

$x \qquad \le \qquad 7$

83. $4x - 2 < 7x + 16$

$\underline{-7x \qquad\ -7x}$

$-3x - 2 < \qquad 16$

$\underline{\ +2 \qquad\quad +2}$

$-3x \qquad < \qquad 18$

$\dfrac{-3x}{-3} > \dfrac{18}{-3}$

$x > -6$

Chapter 3   Polynomials

1. $(x^2)^3 = x^{2\cdot3} = x^6$

3. $(m^4)^4 = m^{4\cdot4} = m^{16}$

5. $(2^4)^2 = 2^{4\cdot2} = 2^8$

7. $(5^3)^5 = 5^{3\cdot5} = 5^{15}$

9. $(3x)^3 = 3^3 \cdot x^3 = 27x^3$

11. $(2xy)^4 = 2^4 \cdot x^4 \cdot y^4 = 16x^4y^4$

13. $5(3ab)^3 = 5 \cdot 3^3 \cdot (ab)^3 = 135a^3b^3$

15. $\left(\dfrac{3}{4}\right)^2 = \dfrac{3^2}{4^2} = \dfrac{9}{16}$

17. $\left(\dfrac{x}{5}\right)^3 = \dfrac{x^3}{5^3} = \dfrac{x^3}{125}$

19. $(2x^2)^4 = 2^4 \cdot (x^2)^4 = 16x^8$

21. $(a^8b^6)^2 = (a^8)^2 \cdot (b^6)^2 = a^{16}b^{12}$

23. $(4x^2y)^3 = 4^3 \cdot \left(x^2\right)^3 y^3 = 64x^6y^3$

25. $(3m^2)^4(m^3)^2 = 81m^8 \cdot m^6 = 81m^{14}$

27. $\dfrac{(x^4)^3}{x^2} = \dfrac{x^{12}}{x^2} = x^{10}$

29. $\dfrac{(s^3)^2(s^2)^3}{(s^5)^2} = \dfrac{s^6 \cdot s^6}{s^{10}} = \dfrac{s^{12}}{s^{10}} = s^2$

31. $\left(\dfrac{m^3}{n^2}\right)^3 = \dfrac{\left(m^3\right)^3}{\left(n^2\right)^3} = \dfrac{m^9}{n^6}$

33. $\left(\dfrac{a^3b^2}{c^4}\right)^2 = \dfrac{\left(a^3\right)^2\left(b^2\right)^2}{\left(c^4\right)^2} = \dfrac{a^6b^4}{c^8}$

35. Polynomial (with a single term)

37. Polynomial

39. Polynomial (with a single term)

41. Not a polynomial because $\dfrac{3+x}{x^2}$ is not a term.

43. Terms: $2x^2,\ -3x$
    Coefficients: 2, –3

45. Terms: $4x^3,\ -3x,\ 2$
    Coefficients: 4, –3, 2

47. Binomial because there are two terms

49. Trinomial because there are three terms

51. Not classified

53. Monomial because there is one term

55. Not a polynomial because $\dfrac{3}{x^2}$ is not a term

57. $4x^5 - 3x^2$: 5th degree

59. $-5x^9 + 7x^7 + 4x^3$: 9th degree

61. $4x$: 1st degree

63. $x^6 - 3x^5 + 5x^2 - 7$: 6th degree

65. $x = 1$: $6x + 1 = 6(1) + 1 = 7$
    $x = -1$: $6x + 1 = 6(-1) + 1 = -5$

67. $x = 2$: $x^3 - 2x = (2)^3 - 2(2) = 4$
    $x = -2$: $x^3 - 2x = (-2)^3 - 2(-2) = -4$

69. $x = 4$: $3x^2 + 4x - 2 = 3(4)^2 + 4(4) - 2$
    $\qquad\qquad\qquad = 48 + 16 - 2$
    $\qquad\qquad\qquad = 62$
    $x = -4$: $3x^2 + 4x - 2 = 3(-4)^2 + 4(-4) - 2$
    $\qquad\qquad\qquad = 48 - 16 - 2$
    $\qquad\qquad\qquad = 30$

71. $x = 1$: $-x^2 - 2x + 3 = -(1)^2 - 2(1) + 3$
    $\qquad\qquad\qquad = -1 - 2 + 3$
    $\qquad\qquad\qquad = 0$
    $x = -3$: $-x^2 - 2x + 3 = -(-3)^2 - 2(-3) + 3$
    $\qquad\qquad\qquad = -9 + 6 + 3$
    $\qquad\qquad\qquad = 0$

73. Always true

75. Sometimes true; The degree of $x^2 + 2x + 1$ is 2 but the degree of $x^3 + x + 1$ is 3.

77. Sometimes true; A polynomial can have any number of terms.

79. $x^{12} = (x^2)^6$

81. $a^{16} = (a^2)^8$

83. $2^{12} = (2^3)^4 = 8^4$
$2^{18} = (2^3)^6 = 8^6$
$(2^5)^3 = (2^3)^5 = 8^5$
$(2^7)^6 = (2^6)^7 = [(2^3)^2]^7 = (8^2)^7 = 8^{14}$

85. $-8x^6 y^9 z^{15} = (-2x^2 y^3 x^5)^3$

87. **a.** $105 = 35 \cdot 3$ so there are three doublings and $[(1.02)^{35}]^3 = (1.02)^{105} \approx 8$. The population will be 8 times as large.

   b. $3.8 \cdot 8 = 30.4$. Their population will be 30.4 billion.

89. Writing exercise

91. $P(1) = (1)^3 - 2(1)^2 + 5 = 4$

93. $Q(2) = 2(2)^2 + 3 = 11$

95. $P(3) = (3)^3 - 2(3)^2 + 5 = 14$

97. $P(0) = (0)^3 - 2(0)^2 + 5 = 5$

99. $P(2) + Q(-1) = [(2)^3 - 2(2)^2 + 5] + [2(-1)^2 + 3]$
$= 5 + 5$
$= 10$

101. $P(3) - Q(-3) \div Q(0)$
$= [(3)^3 - 2(3)^2 + 5] - \{[2(-3)^2 + 3] \div [2(0)^2 + 3]\}$
$= 14 - (21 \div 3)$
$= 7$

103. $|Q(4)| - |P(4)| = \left|2(4)^2 + 3\right| - \left|(4)^3 - 2(4)^2 + 5\right|$
$= |35| - |37|$
$= -2$

105. Cost $= 3y + 20$; $3(50) + 20 = 170$. The cost of typing 50 pages is $170.

107. Revenue $= 3(12)^2 - 95 = 337$. The revenue is $337. The average revenue per pair of shoes is $\frac{337}{12} = \$28.08$.

109. Writing exercise

Section 3.2

1. $4^0 = 1$

3. $(-29)^0 = 1$

5. $(x^3 y^2)^0 = 1$

7. $11x^0 = 11 \cdot 1 = 11$

9. $(-3p^6 q^8)^0 = 1$

11. $b^{-8} = \dfrac{1}{b^8}$

13. $3^{-4} = \dfrac{1}{3^4} = \dfrac{1}{81}$

15. $5^{-2} = \dfrac{1}{5^2} = \dfrac{1}{25}$

17. $10^{-4} = \dfrac{1}{10^4} = \dfrac{1}{10,000}$

19. $5x^{-1} = \dfrac{5}{x}$

21. $(5x)^{-1} = \dfrac{1}{5x}$

23. $-2x^{-5} = -2 \cdot \dfrac{1}{x^5} = -\dfrac{2}{x^5}$

25. $(-2x)^{-5} = \dfrac{1}{(-2x)^5} = -\dfrac{1}{32x^5}$

27. $a^5 a^3 = a^{5+3} = a^8$

29. $x^8 x^{-2} = x^{8+(-2)} = x^6$

31. $b^7 b^{-11} = b^{-4} = \dfrac{1}{b^4}$

33. $x^0 x^5 = 1 \cdot x^5 = x^5$

35. $\dfrac{a^8}{a^5} = a^{8-5} = a^3$

37. $\dfrac{x^7}{x^9} = x^{7-9} = x^{-2} = \dfrac{1}{x^2}$

39. $\dfrac{r^{-3}}{r^5} = r^{-3}r^{-5} = r^{-8} = \dfrac{1}{r^8}$

41. $\dfrac{x^{-4}}{x^{-5}} = x^{-4}x^5 = x$

43. $\dfrac{m^5 n^{-3}}{m^{-4}n^5} = m^{5-(-4)}n^{(-3-5)} = m^9 n^{-8} = \dfrac{m^9}{n^8}$

45. $(2a^{-3})^4 = 16a^{-12} = \dfrac{16}{a^{12}}$

47. $(x^{-2}y^3)^{-2} = x^4 y^{-6} = \dfrac{x^4}{y^6}$

49. $\dfrac{(r^{-2})^3}{r^{-4}} = \dfrac{r^{-6}}{r^{-4}} = r^{-6-(-4)} = r^{-2} = \dfrac{1}{r^2}$

51. $\dfrac{(x^{-3})^3}{(x^4)^{-2}} = \dfrac{x^{-9}}{x^{-8}} = x^{-9-(-8)} = x^{-1} = \dfrac{1}{x}$

53. $\dfrac{(a^{-3})^2(a^4)}{(a^{-3})^{-3}} = \dfrac{a^{-2}}{a^9} = a^{-2-9} = a^{-11} = \dfrac{1}{a^{11}}$

55. $93,000,000 = 9.3 \times 10^7 \, \text{mi}$

57. $130,000,000,000 = 1.3 \times 10^{11} \, \text{cm}$

59. $30 - 2 = 28$ zeros

61. $8 \times 10^{-3} = 0.008$

63. $2.8 \times 10^{-5} = 0.000028$

65. $0.0005 = 5 \times 10^{-4}$

67. $0.00037 = 3.7 \times 10^{-4}$

69. $(4 \times 10^{-3})(2 \times 10^{-5}) = 4 \times 2 \times 10^{-3} \times 10^{-5}$
$$= 8 \times 10^{-8}$$

71. $\dfrac{9 \times 10^3}{3 \times 10^{-2}} = \dfrac{9}{3} \times 10^3 \times 10^2 = 3 \times 10^5$

73. $(2 \times 10^5)(4 \times 10^4) = 2 \times 4 \times 10^5 \times 10^4$
$$= 8 \times 10^9$$

75. $\dfrac{6 \times 10^9}{3 \times 10^7} = \left(\dfrac{6}{3}\right) \times 10^9 \times 10^{-7} = 2 \times 10^2$

77. $\dfrac{(3.3 \times 10^{15})(6 \times 10^{15})}{(1.1 \times 10^8)(3 \times 10^6)}$
$$= \left(\dfrac{3.3}{1.1}\right) \times 10^{15} \times 10^{-8} \cdot \left(\dfrac{6}{3}\right) \times 10^{15} \times 10^{-6}$$
$$= 3 \times 10^7 \cdot 2 \times 10^9$$
$$= 6 \times 10^{16}$$

79. $P = 4 \times 2^{(1960-1975)/35} \approx 2.97$. Earth's population in 1960 was approximately 2.97 billion.

81. $P = 250 \times 2^{(1960-1990)/66} \approx 182.44$. The U.S. population in 1960 was approximately 182 million.

83. $\left(6.6 \times 10^{17}\, \text{m}\right)\left(\dfrac{1 \text{ year}}{1 \times 10^{16}\, \text{m}}\right) = 6.6 \times 10^{17} \times 10^{-16}$
$$= 6.6 \times 10^1$$
$$= 66 \text{ years}$$

85. $15,500 \times 10^{19} = 1.55 \times 10^{23}$ L of water on Earth
$$\dfrac{1.55 \times 10^{23}}{6 \times 10^9} \approx (0.2583) \times 10^{23} \times 10^{-9}$$
$$\approx 0.258 \times 10^{14}$$
$$= 2.58 \times 10^{13} \text{ L per person}$$

87. $\left(\dfrac{2.6 \times 10^6 \, \text{L}}{\text{person}}\right)(3.2 \times 10^8 \text{ people}) = 8.32 \times 10^{14} \text{ L}$

Section 3.3

1. $(6a - 5) + (3a + 9) = 9a + 4$

3. $(8b^2 - 11b) + (5b^2 - 7b) = 13b^2 - 18b$

5. $(3x^2 - 2x) + (-5x^2 + 2x) = -2x^2$

7. $(2x^2 + 5x - 3) + (3x^2 - 7x + 4) = 5x^2 - 2x + 1$

9. $(2b^2 + 8) + (5b + 8) = 2b^2 + 5b + 16$

11. $(8y^3 - 5y^2) + (5y^2 - 2y) = 8y^3 - 2y$

13. $(2a^2 - 4a^3) + (3a^3 + 2a^2) = -a^3 + 4a^2$

15. $(4x^2 - 2 + 7x) + (5 - 8x - 6x^2) = -2x^2 - x + 3$

17. $-(2a + 3b) = -2a - 3b$

19. $5a - (2b - 3c) = 5a - 2b + 3c$

21. $9r - (3r + 5s) = 6r - 5s$

23. $5p - (-3p + 2q) = 8p - 2q$

25. $(2x - 3) - (x + 4) = x - 7$

27. $(4m^2 - 5m) - (3m^2 - 2m) = m^2 - 3m$

29. $(4y^2 + 5y) - (6y^2 + 5y) = -2y^2$

31. $(3x^2 - 5x - 2) - (x^2 - 4x - 3) = 2x^2 - x + 1$

33. $(8a^2 - 9a) - (3a + 7) = 8a^2 - 12a - 7$

35. $(5b - 2b^2) - (4b^2 - 3b) = -6b^2 + 8b$

37. $(3x^2 - 8x + 7) - (x^2 - 5 - 8x) = 2x^2 + 12$

39. $[(4b - 2) + (5b + 3)] - (3b + 2) = 9b + 1 - 3b - 2$
$$= 6b - 1$$

41. $[(x^2 + 5x - 2) + (2x^2 + 7x - 8)] - (3x^2 + 2x - 1)$
$$= (3x^2 + 12x - 10) - 3x^2 - 2x + 1$$
$$= 10x - 9$$

43. $[(4x^2 - 5) + (2x - 7)] - (2x^2 - 3x)$
$$= (4x^2 + 2x - 12) - 2x^2 + 3x$$
$$= 2x^2 + 5x - 12$$

45. $(2y^2 - 8y) - [(3y^2 - 3y) + (5y^2 + 3y)]$
$$= (2y^2 - 8y) - 8y^2$$
$$= -6y^2 - 8y$$

47.
$$\begin{array}{r} 2w^2 + 7 \\ 3w - 5 \\ \underline{4w^2 - 5w } \\ 6w^2 - 2w + 2 \end{array}$$

49.
$$\begin{array}{r} 3x^2 + 3x - 4 \\ 4x^2 - 3x - 3 \\ \underline{2x^2 - x + 7} \\ 9x^2 - x \end{array}$$

51.
$$\begin{array}{cc} 5a^2 + 3a & 5a^2 + 3a \\ \underline{(-)3a^2 - 2a} & \underline{-3a^2 + 2a} \\ & 2a^2 + 5a \end{array}$$

53.
$$\begin{array}{cc} 8x^2 - 5x + 7 & 8x^2 - 5x + 7 \\ \underline{(-)5x^2 - 6x + 7} & \underline{-5x^2 + 6x - 7} \\ & 3x^2 + x \end{array}$$

55.
$$\begin{array}{cc} 8x^2 - 9 & 8x^2 - 9 \\ \underline{(-)5x^2 - 3x } & \underline{-5x^2 + 3x } \\ & 3x^2 + 3x - 9 \end{array}$$

57. $[(9x^2 - 3x + 5) - (3x^2 + 2x - 1)] - (x^2 - 2x - 3)$
$$= (6x^2 - 5x + 6) - x^2 + 2x + 3$$
$$= 5x^2 - 3x + 9$$

59. $3ax^4 - 5x^3 + x^2 - cx + 2 = 9x^4 - bx^3 + x^2 - 2d$
$$3a = 9 \quad -5 = -b \quad -c = 0 \quad 2 = -2d$$
$$a = 3 \quad\quad b = 5 \quad\quad c = 0 \quad d = -1$$

61. Perimeter $= 2l + 2w$
$$= 2(8x + 9) + 2(6x - 7) = 28x + 4$$

63. Profit $= R - C$
$$= (90x - x^2) - (150 + 25x)$$
$$= -x^2 + 65x - 150$$

Section 3.4

1. $(5x^2)(3x^3) = (5 \cdot 3)(x^2 \cdot x^3)$
$$= 15x^5$$

3. $(-2b^2)(14b^8) = (-2 \cdot 14)(b^2 \cdot b^8)$
$$= -28b^{10}$$

5. $(-10p^6)(-4p^7) = (-10)(-4)(p^6 \cdot p^7)$
$= 40p^{13}$

7. $(4m^5)(-3m) = (4)(-3)(m^5 \cdot m)$
$= -12m^6$

9. $(4x^3y^2)(8x^2y) = (4 \cdot 8)(x^3 \cdot x^2)(y^2 \cdot y)$
$= 32x^5y^3$

11. $(-3m^5n^2)(2m^4n) = (-3)(2)(m^5 \cdot m^4)(n^2 \cdot n)$
$= -6m^9n^3$

13. $5(2x + 6) = 5(2x) + 5(6)$
$= 10x + 30$

15. $3a(4a + 5) = 3a(4a) + 3a(5)$
$= 12a^2 + 15a$

17. $3s^2(4s^2 - 7s) = 3s^2(4s^2) - 3s^2(7s)$
$= 12s^4 - 21s^3$

19. $2x(4x^2 - 2x + 1) = 2x(4x^2) - 2x(2x) + 2x(1)$
$= 8x^3 - 4x^2 + 2x$

21. $3xy(2x^2y + xy^2 + 5xy)$
$= 3xy(2x^2y) + 3xy(xy^2) + 3xy(5xy)$
$= 6x^3y^2 + 3x^2y^3 + 15x^2y^2$

23. $6m^2n(3m^2n - 2mn + mn^2)$
$= 6m^2n(3m^2n) - 6m^2n(2mn) + 6m^2n(mn^2)$
$= 18m^4n^2 - 12m^3n^2 + 6m^3n^3$

25. $(x + 3)(x + 2) = x^2 + 2x + 3x + 6$
$= x^2 + 5x + 6$

27. $(m - 5)(m - 9) = m^2 - 9m - 5m + 45$
$= m^2 - 14m + 45$

29. $(p - 8)(p + 7) = p^2 + 7p - 8p - 56$
$= p^2 - p - 56$

31. $(w + 10)(w + 20) = w^2 + 20w + 10w + 200$
$= w^2 + 30w + 200$

33. $(3x - 5)(x - 8) = 3x^2 - 24x - 5x + 40$
$= 3x^2 - 29x + 40$

35. $(2x - 3)(3x + 4) = 6x^2 + 8x - 9x - 12$
$= 6x^2 - x - 12$

37. $(3a - b)(4a - 9b) = 12a^2 - 27ab - 4ab + 9b^2$
$= 12a^2 - 31ab + 9b^2$

39. $(3p - 4q)(7p + 5q) = 21p^2 + 15pq - 28pq - 20q^2$
$= 21p^2 - 13pq - 20q^2$

41. $(2x + 5y)(3x + 4y) = 6x^2 + 8xy + 15xy + 20y^2$
$= 6x^2 + 23xy + 20y^2$

43. $(x + 5)^2 = (x + 5)(x + 5)$
$= x^2 + 5x + 5x + 25$
$= x^2 + 10x + 25$

45. $(y - 9)^2 = (y - 9)(y - 9)$
$= y^2 - 9y - 9y + 81$
$= y^2 - 18y + 81$

47. $(6m + n)^2 = (6m + n)(6m + n)$
$= 36m^2 + 6mn + 6mn + n^2$
$= 36m^2 + 12mn + n^2$

49. $(a - 5)(a + 5) = a^2 + 5a - 5a - 25 = a^2 - 25$

51. $(x - 2y)(x + 2y) = x^2 + 2xy - 2xy - 4y^2$
$= x^2 - 4y^2$

53. $(5s + 3t)(5s - 3t) = 25s^2 - 15st + 15st - 9t^2$
$= 25s^2 - 9t^2$

55.
$$
\begin{array}{r}
x + 2 \\
3x + 5 \\
\hline
5x + 10 \\
3x^2 + 6x \\
\hline
3x^2 + 11x + 10
\end{array}
$$

57.
$$
\begin{array}{r}
2m + (-5) \\
3m + 7 \\
\hline
14m - 35 \\
6m^2 + (-15m) \\
\hline
6m^2 - \ \ m - 35
\end{array}
$$

59.

$$
\begin{array}{r}
3x + 4y \\
5x + (-2y) \\
\hline
-6xy + (-8y^2) \\
15x^2 + 20xy \\
\hline
15x^2 + 14xy - 8y^2
\end{array}
$$

61.

$$
\begin{array}{r}
a^2 + 3ab - b^2 \\
a^2 - 5ab + b^2 \\
\hline
a^2b^2 + 3ab^3 - b^4 \\
-5a^3b - 15a^2b^2 + 5ab^3 \\
a^4 + 3a^3b - a^2b^2 \\
\hline
a^4 - 2a^3b - 15a^2b^2 + 8ab^3 - b^4
\end{array}
$$

63.

$$
\begin{array}{r}
x^2 + 2xy + 4y^2 \\
x + (-2y) \\
\hline
-2x^2y + (-4xy^2) + (-8y^3) \\
x^3 + 2x^2y + 4xy^2 \\
\hline
x^3 \qquad\qquad - 8y^3
\end{array}
$$

65.

$$
\begin{array}{r}
9a^2 - 12ab + 16b^2 \\
\cdot\ 3a + 4b \\
\hline
36a^2b - 48ab^2 + 64b^3 \\
27a^3 + (-36a^2b) + 48ab^2 \\
\hline
27a^3 \qquad\qquad\qquad + 64b^3
\end{array}
$$

67. $2x(3x-2)(4x+1) = 2x(12x^2 + 3x - 8x - 2)$

$\qquad\qquad\qquad\quad = 2x(12x^2 - 5x - 2)$

$\qquad\qquad\qquad\quad = 24x^3 - 10x^2 - 4x$

69. $5a(4a-3)(4a+3) = 5a(16a^2 + 12a - 12a - 9)$

$\qquad\qquad\qquad\quad = 5a(16a^2 - 9)$

$\qquad\qquad\qquad\quad = 80a^3 - 45a$

71. $3s(5s-2)(4s-1) = 3s(20s^2 - 13s + 2)$

$\qquad\qquad\qquad\quad = 60s^3 - 39s^2 + 6s$

73. $(x-2)(x+1)(x-3)$

$\qquad = (x-2)(x^2 - 2x - 3)$

$\qquad = x^3 - 2x^2 - 3x - 2x^2 + 4x + 6$

$\qquad = x^3 - 4x^2 + x + 6$

75. $(a-1)^3 = (a-1)(a-1)(a-1)$

$\qquad\quad = (a-1)(a^2 - 2a + 1)$

$\qquad\quad = a^3 - 2a^2 + a - a^2 + 2a - 1$

$\qquad\quad = a^3 - 3a^2 + 3a - 1$

77. $\left(\dfrac{x}{2} + \dfrac{2}{3}\right)\left(\dfrac{2x}{3} - \dfrac{2}{5}\right) = \dfrac{x^2}{3} - \dfrac{x}{5} + \dfrac{4x}{9} - \dfrac{4}{15}$

$\qquad\qquad\qquad\qquad = \dfrac{1}{3}x^2 + \dfrac{11}{45}x - \dfrac{4}{15}$

79. $[x+(y-2)][x-(y-2)] = x^2 - (y-2)^2$

$\qquad\qquad\qquad\qquad = x^2 - (y^2 - 4y + 4)$

$\qquad\qquad\qquad\qquad = x^2 - y^2 + 4y - 4$

81. $(x+y)^2 = x^2 + 2xy + y^2$

$\qquad\quad \neq x^2 + y^2$: False

83. $(x+y)^2 = x^2 + 2xy + y^2$: True

85. Area $= (3x+5)(2x-7)$

$\qquad\quad = 6x^2 - 11x - 35$ cm^2

87. Revenue $= x(2x-10) = 2x^2 - 10x$

89. Group exercise

Section 3.5

1. $(x+5)^2 = x^2 + 2(5x) + 5^2 = x^2 + 10x + 25$

3. $(w-6)^2 = w^2 - 2(6w) + 6^2 = w^2 - 12w + 36$

5. $(z+12)^2 = z^2 + 2(12z) + 12^2 = z^2 + 24z + 144$

7. $(2a-1)^2 = (2a)^2 - 2(2a) + 1^2 = 4a^2 - 4a + 1$

9. $(6m+1)^2 = (6m)^2 + 2(6m) + 1^2$

$\qquad\qquad = 36m^2 + 12m + 1$

11. $(3x-y)^2 = (3x)^2 - 2(3xy) + y^2$

$\qquad\qquad = 9x^2 - 6xy + y^2$

13. $(2r+5s)^2 = (2r)^2 + 2(2r)(5s) + (5s)^2$

$\qquad\qquad = 4r^2 + 20rs + 25s^2$

15. $(8a-9b)^2 = (8a)^2 - 2(8a)(9b) + (9b)^2$

$\qquad\qquad = 64a^2 - 144ab + 81b^2$

17. $\left(x+\dfrac{1}{2}\right)^2 = x^2 + 2\left(\dfrac{1}{2}x\right) + \left(\dfrac{1}{2}\right)^2$
$$= x^2 + x + \dfrac{1}{4}$$

19. $(x-6)(x+6) = x^2 - 6^2 = x^2 - 36$

21. $(m+12)(m-12) = m^2 - 12^2 = m^2 - 144$

23. $\left(x-\dfrac{1}{2}\right)\left(x+\dfrac{1}{2}\right) = x^2 - \left(\dfrac{1}{2}\right)^2 = x^2 - \dfrac{1}{4}$

25. $(p-0.4)(p+0.4) = p^2 - (0.4)^2 = p^2 - 0.16$

27. $(a-3b)(a+3b) = a^2 - (3b)^2 = a^2 - 9b^2$

29. $(4r-s)(4r+s) = (4r)^2 - s^2 = 16r^2 - s^2$

31. $(8w+5z)(8w-5z) = (8w)^2 - (5z)^2$
$$= 64w^2 - 25z^2$$

33. $(5x-9y)(5x+9y) = (5x)^2 - (9y)^2$
$$= 25x^2 - 81y^2$$

35. $x(x-2)(x+2) = x(x^2 - 4) = x^3 - 4x$

37. $2s(s-3r)(s+3r) = 2s(s^2 - 9r^2) = 2s^3 - 18sr^2$

39. $5r(r+3)^2 = 5r(r^2 + 6r + 9)$
$$= 5r^3 + 30r^2 + 45r$$

41. The product of 6 more than a number and 6 less than that number $= (x+6)(x-6)$
$$= x^2 - 36.$$

43. The square of 4 less than a number
$$= (x-4)^2$$
$$= x^2 - 8x + 16$$

45. $(49)(51) = (50-1)(50+1) = 2500 - 1 = 2499$

47. $(34)(26) = (30+4)(30-4) = 900 - 16 = 884$

49. $(55)(65) = (60-5)(60+5) = 3600 - 25 = 3575$

51. $(5x-4)^2 = 25x^2 - 40x + 16$ trees

53. Writing exercise

55. Challenge exercise

Section 3.6

1. $\dfrac{18x^6}{9x^2} = 2x^{6-2} = 2x^4$

3. $\dfrac{35m^3 n^2}{7mn^2} = 5m^{3-1}n^{2-2} = 5m^2$

5. $\dfrac{3a+6}{3} = \dfrac{3a}{3} + \dfrac{6}{3} = a + 2$

7. $\dfrac{9b^2 - 12}{3} = \dfrac{9b^2}{3} - \dfrac{12}{3} = 3b^2 - 4$

9. $\dfrac{16a^3 - 24a^2}{4a} = \dfrac{16a^3}{4a} - \dfrac{24a^2}{4a} = 4a^2 - 6a$

11. $\dfrac{12m^2 + 6m}{-3m} = \dfrac{12m^2}{-3m} + \dfrac{6m}{-3m} = -4m - 2$

13. $\dfrac{18a^4 + 12a^3 - 6a^2}{6a} = \dfrac{18a^4}{6a} + \dfrac{12a^3}{6a} - \dfrac{6a^2}{6a}$
$$= 3a^3 + 2a^2 - a$$

15. $\dfrac{20x^4 y^2 - 15x^2 y^3 + 10x^3 y}{5x^2 y} = \dfrac{20x^4 y^2}{5x^2 y} - \dfrac{15x^2 y^3}{5x^2 y}$
$$= 4x^2 y - 3y^2 + 2x$$

17. Since the divisor is a binomial, use long division.

$$\begin{array}{r} x+3 \\ x+2\overline{\smash{)}x^2 + 5x + 6} \\ \underline{x^2 + 2x} \\ 3x + 6 \\ \underline{3x + 6} \\ 0 \end{array}$$

$$\dfrac{x^2 + 5x + 6}{x+2} = x + 3$$

19. Since the divisor is a binomial, use long division.

$$
\begin{array}{r}
x-5 \\
x+4\overline{\smash{)}x^2-x-20} \\
\underline{x^2+4x} \\
-5x-20 \\
\underline{-5x-20} \\
0
\end{array}
$$

$$\frac{x^2-x-20}{x+4}=x-5$$

21. Since the divisor is a binomial, use long division.

$$
\begin{array}{r}
x+3 \\
2x-1\overline{\smash{)}2x^2+5x-3} \\
\underline{2x^2-x} \\
6x-3 \\
\underline{6x-3} \\
0
\end{array}
$$

$$\frac{2x^2+5x-3}{2x-1}=x+3$$

23. Since the divisor is a binomial, use long division.

$$
\begin{array}{r}
2x+3 \\
x-3\overline{\smash{)}2x^2-3x-5} \\
\underline{2x^2-6x} \\
3x-5 \\
\underline{3x-9} \\
4
\end{array}
$$

$$\frac{2x^2-3x-5}{x-3}=2x+3+\frac{4}{x-3}$$

25. Since the divisor is a binomial, use long division.

$$
\begin{array}{r}
4x+2 \\
x-5\overline{\smash{)}4x^2-18x-15} \\
\underline{4x^2-20x} \\
2x-15 \\
\underline{2x-10} \\
-5
\end{array}
$$

$$\frac{4x^2-18x-15}{x-5}=4x+2+\frac{-5}{x-5}$$

27. Since the divisor is a binomial, use long division.

$$
\begin{array}{r}
2x+3 \\
3x-5\overline{\smash{)}6x^2-x-10} \\
\underline{6x^2-10x} \\
9x-10 \\
\underline{9x-15} \\
5
\end{array}
$$

$$\frac{6x^2-x-10}{3x-5}=2x+3+\frac{5}{3x-5}$$

29. Since the divisor is a binomial, use long division.

$$
\begin{array}{r}
x^2-x-2 \\
x+2\overline{\smash{)}x^3+x^2-4x-4} \\
\underline{x^3+2x^2} \\
-x^2-4x \\
\underline{-x^2-2x} \\
-2x-4 \\
\underline{-2x-4} \\
0
\end{array}
$$

$$\frac{x^3+x^2-4x-4}{x+2}=x^2-x-2$$

31. Since the divisor is a binomial, use long division.

$$
\begin{array}{r}
x^2+2x+3 \\
4x-1\overline{\smash{)}4x^3+7x^2+10x+5} \\
\underline{4x^3-x^2} \\
8x^2+10x \\
\underline{8x^2-2x} \\
12x+5 \\
\underline{12x-3} \\
8
\end{array}
$$

$$\frac{4x^3+7x^2+10x+5}{4x-1}=x^2+2x+3+\frac{8}{4x-1}$$

33. Since the divisor is a binomial, use long division. The dividend $x^3 - x^2 + 5$ is missing a term in x, so write $0 \cdot x$.

$$
\begin{array}{r}
x^2 + x + 2 \\
x - 2 \overline{\smash{\big)}\, x^3 - x^2 + 0x + 5} \\
\underline{x^3 - 2x^2} \\
x^2 + 0x \\
\underline{x^2 - 2x} \\
2x + 5 \\
\underline{2x - 4} \\
9
\end{array}
$$

$$\frac{x^3 - x^2 + 5}{x - 2} = x^2 + x + 2 + \frac{9}{x - 2}$$

35. Since the divisor is a binomial, use long division. The dividend $25x^2 + x$ is missing terms in x^2 and x^0, so write $0x^2$ and 0.

$$
\begin{array}{r}
5x^2 + 2x + 1 \\
5x - 2 \overline{\smash{\big)}\, 25x^3 + 0x^2 + x + 0} \\
\underline{25x^3 - 10x^2} \\
10x^2 + x \\
\underline{10x^2 - 4x} \\
5x + 0 \\
\underline{5x - 2} \\
2
\end{array}
$$

$$\frac{25x^2 + x}{5x - 2} = 5x^2 + 2x + 1 + \frac{2}{5x - 2}$$

37. Since the divisor is a binomial, use long division. Rearrange the dividend in descending-exponent form.

$$
\begin{array}{r}
x^2 + 4x + 5 \\
x - 2 \overline{\smash{\big)}\, x^3 + 2x^2 - 3x - 8} \\
\underline{x^3 - 2x^2} \\
4x^2 - 3x \\
\underline{4x^2 - 8x} \\
5x - 8 \\
\underline{5x - 10} \\
2
\end{array}
$$

$$\frac{2x^2 - 8 - 3x + x^3}{x - 2} = x^2 + 4x + 5 + \frac{2}{x - 2}$$

39. Since the divisor is a binomial, use long division. The dividend $x^4 - 1$ is "missing" terms in x^3, x^2, and x, so write $0x^3 + 0x^2 + 0x$.

$$
\begin{array}{r}
x^3 + x^2 + x + 1 \\
x - 1 \overline{\smash{\big)}\, x^4 + 0x^3 + 0x^2 + 0x - 1} \\
\underline{x^4 - x^3} \\
x^3 + 0x^2 \\
\underline{x^3 - x^2} \\
x^2 + 0x \\
\underline{x^2 - x} \\
x - 1 \\
\underline{x - 1} \\
0
\end{array}
$$

$$\frac{x^4 - 1}{x - 1} = x^3 + x^2 + x + 1$$

41. Since the divisor is a binomial, use long division.

$$
\begin{array}{r}
x - 3 \\
x^2 - 1 \overline{\smash{\big)}\, x^3 - 3x^2 - x + 3} \\
\underline{x^3 - x} \\
-3x^2 + 3 \\
\underline{-3x^2 + 3} \\
0
\end{array}
$$

$$\frac{x^3 - 3x^2 - x + 3}{x^2 - 1} = x - 3$$

43. Since the divisor is a binomial, use long division. The dividend is missing terms in x^3 and x, so write $0x^3$ and $0x$.

$$
\begin{array}{r}
x^2 - 1 \\
x^2 + 3 \overline{\smash{\big)}\, x^4 + 0x^3 + 2x^2 + 0x - 2} \\
\underline{x^4 + 3x^2} \\
-x^2 - 2 \\
\underline{-x^2 - 3} \\
1
\end{array}
$$

$$\frac{x^4 + 2x^2 - 2}{x^2 + 3} = x^2 - 1 + \frac{1}{x^2 + 3}$$

45. Since the divisor is a binomial, use long division. The dividend is missing terms in y^2 and y, so write $0y^2$ and $0y$.

$$
\begin{array}{r}
y^2 - y + 1 \\
y+1\overline{\smash{)}\,y^3 + 0y^2 + 0y + 1} \\
\underline{y^3 + y^2} \\
-y^2 + 0y \\
\underline{-y^2 - y} \\
y + 1 \\
\underline{y + 1} \\
0
\end{array}
$$

$$\frac{y^3 + 1}{y + 1} = y^2 - y + 1$$

47. Since the divisor is a binomial, use long division. The dividend is missing terms in x^3, x^2, and x, so write $0x^3$, $0x^2$, and $0x$.

$$
\begin{array}{r}
x^2 + 1 \\
x^2-1\overline{\smash{)}\,x^4 + 0x^3 + 0x^2 + 0x - 1} \\
\underline{x^4 \qquad - x^2} \\
x^2 \qquad - 1 \\
\underline{x^2 \qquad - 1} \\
0
\end{array}
$$

$$\frac{x^4 - 1}{x^2 - 1} = x^2 + 1$$

49. $\dfrac{y^2 - y + c}{y + 1} = y - 2$ can be written as

$$y^2 - y + c = (y - 2)(y + 1)$$
$$y^2 - y + c = y^2 - y - 2$$

Therefore, $c = -2$.

51. Writing exercise

53. a.

$$
\begin{array}{r}
x + 1 \\
x-1\overline{\smash{)}\,x^2 + 0x - 1} \\
\underline{x^2 - x} \\
x - 1 \\
\underline{x - 1} \\
0
\end{array}
$$

$$\frac{x^2 - 1}{x - 1} = x + 1$$

b.

$$
\begin{array}{r}
x^2 + x + 1 \\
x-1\overline{\smash{)}\,x^3 + 0x^2 + 0x - 1} \\
\underline{x^2 - x^2} \\
x^2 + 0x \\
\underline{x^2 - x} \\
x - 1 \\
\underline{x - 1} \\
0
\end{array}
$$

$$\frac{x^3 - 1}{x - 1} = x^2 + x + 1$$

c.

$$
\begin{array}{r}
x^3 + x^2 + x + 1 \\
x-1\overline{\smash{)}\,x^4 + 0x^3 + 0x^2 + 0x - 1} \\
\underline{x^4 - x^3} \\
x^3 + 0x^2 \\
\underline{x^3 - x^2} \\
x^2 + 0x \\
\underline{x^2 - x} \\
x - 1 \\
\underline{x - 1} \\
0
\end{array}
$$

$$\frac{x^4 - 1}{x - 1} = x^3 + x^2 + x + 1$$

d. In each problem (a), (b), and (c) the quotient is a polynomial of degree one less than the degree of the dividend. The quotient has no missing terms and all of its coefficients are 1. Therefore, it would seem like

$$\frac{x^{50} - 1}{x - 1} = x^{49} + x^{48} + \cdots + x + 1.$$

Summary Exercises for Chapter 3

1. $\dfrac{x^{10}}{x^3} = x^{10-3} = x^7$

3. $\dfrac{x^2 \cdot x^3}{x^4} = \dfrac{x^5}{x^4} = x$

5. $\dfrac{18p^7}{9p^5} = 2p^{7-5} = 2p^2$

7. $\dfrac{30m^7n^5}{6m^2n^3} = 5m^{7-2}n^{5-3} = 5m^5n^2$

9. $\dfrac{48p^5q^3}{6p^3q} = 8p^{5-3}q^{3-1} = 8p^2q^2$

11. $(2ab)^2 = 2^2a^2b^2 = 4a^2b^2$

13. $(2x^2y^2)^3(3x^3y)^2 = (8x^6y^6)(9x^6y^2) = 72x^{12}y^8$

15. $\dfrac{(x^5)^2}{(x^3)^3} = \dfrac{x^{10}}{x^9} = x$

17. $(y^3)^2(3y^2)^3 = (y^6)(27y^6) = 27y^{12}$

19. $5(-1)+1 = -5+1 = -4$

21. $-(6)^2 + 3(6) - 1 = -36 + 18 - 1 = -19$

23. Binomial because there are two terms

25. Trinomial because there are three terms

27. Binomial because there are two terms

29. $9x$; 1st degree

31. $x + 5$: 1st degree

33. $7x^6 + 9x^4 - 3x$; 6th degree

35. $(3a)^0 = 1$

37. $(3a4b)^0 = 1$

39. $3^{-3} = \dfrac{1}{3^3}$

41. $4x^{-4} = 4 \cdot \dfrac{1}{x^4} = \dfrac{4}{x^4}$

43. $m^7m^{-9} = m^{-2} = \dfrac{1}{m^2}$

45. $\dfrac{x^2y^{-3}}{x^{-3}y^2} = x^2y^{-3}x^3y^{-2} = x^5y^{-5} = \dfrac{x^5}{y^5}$

47. $\dfrac{(a^4)^{-3}}{(a^{-2})^{-3}} = \dfrac{a^{-12}}{a^6} = \dfrac{1}{a^{12}a^6} = \dfrac{1}{a^{18}}$

49. $51{,}000 = 5.1 \times 10^4$ cycles/sec

51. $(2.3 \times 10^{-3})(1.4 \times 10^{12})$
$= (2.3 \times 1.4) \times (10^{-3} \times 10^{12})$
$= 3.22 \times 10^9$

53. $\dfrac{(8 \times 10^{23})}{(4 \times 10^6)} = 2 \times 10^{23-6} = 2 \times 10^{17}$

55. $9a^2 - 5a$
$\underline{12a^2 + 3a}$
$21a^2 - 2a$

57. $5y^3 - 3y^2$
$\underline{ 3y^2 + 4y}$
$5y^3 + 4y$

59. $7x^2 - 2x + 3$
$\underline{-2x^2 + 5x + 7}$
$5x^2 + 3x + 10$

61. $[(9x+2)+(-3x-7)]-(5x-3) = 6x-5-5x+3$
$= x - 2$

63. $7w^2 - 5w + 2 - [(16w^2 - 3w) + (8w + 2)]$
$= 7w^2 - 5w + 2 - (16w^2 + 5w + 2)$
$= 7w^2 - 5w + 2 - 16w^2 - 5w - 2$
$= -9w^2 - 10w$

65. $9b^2 - 7$
$\underline{ 8b + 5}$
$9b^2 + 8b - 2$

67. $7x^2 - 5x - 7$
$\underline{-5x^2 + 3x - 2}$
$2x^2 - 2x - 9$

69. $(5a^3)(a^2) = 5a^{3+2} = 5a^5$

71. $(-9p^3)(-6p^2) = (-9)(-6)p^{3+2} = 54p^5$

73. $5(3x - 8) = 5(3x) - 5(8) = 15x - 40$

75. $(-5rs)(2r^2s - 5rs) = -5rs(2r^2s) - 5rs(-5rs)$
$$= -10r^3s^2 + 25r^2s^2$$

77. $(x + 5)(x + 4) = x^2 + 4x + 5x + 20$
$$= x^2 + 9x + 20$$

79. $(a - 7b)(a + 7b) = a^2 - (7b)^2 = a^2 - 49b^2$

81. $(a + 4b)(a + 3b) = a^2 + 3ab + 4ab + 12b^2$
$$= a^2 + 7ab + 12b^2$$

83. $(3x - 5y)(2x - 3y) = 6x^2 - 9xy - 10xy + 15y^2$
$$= 6x^2 - 19xy + 15y^2$$

85. $(y + 2)(y^2 - 2y + 3)$
$$= y^3 - 2y^2 + 3y + 2y^2 - 4y + 6$$
$$= y^3 - y + 6$$

87. $(x - 2)(x^2 + 2x + 4)$
$$= x^3 + 2x^2 + 4x - 2x^2 - 4x - 8$$
$$= x^3 - 8$$

89. $2x(x + 5)(x - 6) = 2x(x^2 - 6x + 5x - 30)$
$$= 2x(x^2 - x - 30)$$
$$= 2x^3 - 2x^2 - 60x$$

91. $(x + 7)^2 = x^2 + 2(x)(7) + 7^2$
$$= x^2 + 14x + 49$$

93. $(2w - 5)^2 = (2w)^2 + 2(2w)(-5) + (-5)^2$
$$= 4w^2 - 20w + 25$$

95. $(a + 7b)^2 = a^2 + 2(a)(7b) + (7b)^2$
$$= a^2 + 14ab + 49b^2$$

97. $(x - 5)(x + 5) = x^2 - 25$

99. $(2m + 3)(2m - 3) = 4m^2 - 9$

101. $(5r - 2s)(5r + 2s) = 25r^2 - 4s^2$

103. $2x(x - 5)^2 = 2x(x^2 + 2(x)(-5) + (-5)^2)$
$$= 2x(x^2 - 10x + 25)$$
$$= 2x^3 - 20x^2 + 50x$$

105. $\dfrac{9a^5}{3a^2} = 3a^{5-2} = 3a^3$

107. $\dfrac{15a - 10}{5} = \dfrac{15a}{5} - \dfrac{10}{5} = 3a - 2$

109. $\dfrac{9r^2s^3 - 18r^3s^2}{-3rs^2} = \dfrac{9r^2s^3}{-3rs^2} - \dfrac{18r^3s^2}{-3rs^2} = -3rs + 6r$

111. Since the divisor is a binomial, use long division.

$$\begin{array}{r} x - 5 \\ x+3 \overline{)\, x^2 - 2x - 15} \\ \underline{x^2 + 3x } \\ -5x - 15 \\ \underline{-5x - 15} \\ 0 \end{array}$$

$$\dfrac{x^2 - 2x - 15}{x + 3} = x - 5$$

113. Since the divisor is a binomial, use long division.

$$\begin{array}{r} x - 3 \\ x-5 \overline{)\, x^2 - 8x + 17} \\ \underline{x^2 - 5x } \\ -3x + 17 \\ \underline{-3x + 15} \\ 2 \end{array}$$

$$\dfrac{x^2 - 8x + 17}{x - 5} = x - 3 + \dfrac{2}{x - 5}$$

115. Since the divisor is a binomial, use long division.

$$\begin{array}{r} x^2 + 2x - 1 \\ 6x+2 \overline{)\, 6x^3 + 14x^2 - 2x - 6} \\ \underline{6x^3 + 2x^2 } \\ 12x^2 - 2x \\ \underline{12x^2 + 4x} \\ -6x - 6 \\ \underline{-6x - 2} \\ -4 \end{array}$$

$$\dfrac{6x^3 + 14x^2 - 2x - 6}{6x + 2} = x^2 + 2x - 1 + \dfrac{-4}{6x + 2}$$

117. Since the divisor is a binomial, use long division. Rearrange the dividend in descending-exponent form.

$$
\begin{array}{r}
x^2 + x + 2 \\
x+2\overline{\smash{\big)}\,x^3 + 3x^2 + 4x + 5} \\
\underline{x^3 + 2x^2} \\
x^2 + 4x \\
\underline{x^2 + 2x} \\
2x + 5 \\
\underline{2x + 4} \\
1
\end{array}
$$

$$
\frac{3x^2 + x^3 + 5 + 4x}{x+2} = x^2 + x + 2 + \frac{1}{x+2}
$$

Chapter 4   Factoring

Exercises 4.1

1. 10, 12
The largest number that is a factor of both is 2. The GCF is 2.

3. 16, 32, 88
The GCF is 8.

5. x^2, x^5
The largest power that divides both terms is x^2. The GCF is x^2.

7. a^3, a^6, a^9
The GCF is a^3.

9. $5x^4$, $10x^5$
The GCF is $5x^4$.

11. $8a^4$, $6a^6$, $10a^{10}$
The GCF is $2a^4$.

13. $9x^2y$, $12xy^2$, $15x^2y^2$
The GCF is $3xy$.

15. $15ab^3$, $10a^2bc$, $25b^2c^3$
The GCF is $5b$.

17. $15a^2bc^2$, $9ab^2c^2$, $6a^2b^2c^2$
The GCF is $3abc^2$.

19. $(x+y)^2$, $(x+y)^3$
$x + y$ is a binomial used as a factor.
The GCF is $(x + y)^2$.

21. $8a + 4 = 4 \cdot 2a + 4 \cdot 1$
$= 4(2a + 1)$

23. $24m - 32n = 8 \cdot 3m + 8(-4n)$
$= 8(3m - 4n)$

25. $12m^2 + 8m = 4m \cdot 3m + 4m \cdot 2$
$= 4m(3m + 2)$

27. $10s^2 + 5s = 5s \cdot 2s + 5s \cdot 1$
$= 5s(2s + 1)$

29. $12x^2 + 24x = 12x \cdot x + 12x \cdot 2$
$= 12x(x + 2)$

31. $15a^3 - 25a^2 = 5a^2 \cdot 3a - 5a^2 \cdot 5$
$= 5a^2(3a - 5)$

33. $6pq + 18p^2q = 6pq \cdot 1 + 6pq \cdot 3p$
$= 6pq(1 + 3p)$

35. $7m^3n - 21mn^3 = 7mn \cdot m^2 - 7mn \cdot 3n^2$
$= 7mn(m^2 - 3n^2)$

37. $6x^2 - 18x + 30 = 6 \cdot x^2 - 6 \cdot 3x + 6 \cdot 5$
$= 6(x^2 - 3x + 5)$

39. $3a^3 + 6a^2 - 12a = 3a \cdot a^2 + 3a \cdot 2a - 3a \cdot 4$
$= 3a(a^2 + 2a - 4)$

41. $6m + 9mn - 15mn^2$
$= 3m \cdot 2 + 3m \cdot 3n - 3m \cdot 5n^2$
$= 3m(2 + 3n - 5n^2)$

43. $10x^2y + 15xy - 5xy^2$
$= 5xy \cdot 2x + 5xy \cdot 3 - 5xy \cdot y$
$= 5xy(2x + 3 - y)$

45. $10r^3s^2 + 25r^2s^2 - 15r^2s^3$
$= 5r^2s^2 \cdot 2r + 5r^2s^2 \cdot 5 - 5r^2s^2 \cdot 3s$
$= 5r^2s^2(2r + 5 - 3s)$

47. $9a^5 - 15a^4 + 21a^3 - 27a$
$= 3a \cdot 3a^4 - 3a \cdot 5a^3 + 3a \cdot 7a^2 - 3a \cdot 9$
$= 3a(3a^4 - 5a^3 + 7a^2 - 9)$

49. $15m^3n^2 - 20m^2n + 35mn^3 - 10mn$
$= 5mn \cdot 3m^2n - 5mn \cdot 4m + 5mn \cdot 7n^2 - 5mn \cdot 2$
$= 5mn(3m^2n - 4m + 7n^2 - 2)$

51. $x(x - 2) + 3(x - 2)$
Notice that the binomial $x - 2$ is a common factor.
$x(x - 2) + 3(x - 2) = (x - 2) \cdot x + (x - 2) \cdot 3$
$= (x - 2)(x + 3)$

53. To find the GCF of the product $(2x - 6)(5x + 10)$, first multiply the factors to get a polynomial.
$(2x - 6)(5x + 10) = 10x^2 - 10x - 60$
$= 10 \cdot x^2 - 10 \cdot x - 10 \cdot 6$
The GCF is 10, or $5 \cdot 2$.

55. To find the GCF of the product
$(2x^3 - 4x)(3x + 6)$, multiply the factors to
get a polynomial.

$(2x^3 - 4x)(3x + 6)$
$= 6x^4 + 12x^3 - 12x^2 - 24x$
$= 6x \cdot x^3 + 6x \cdot 2x^2 - 6x \cdot 2x - 6x \cdot 4$
The GCF is $6x$, or $2x \cdot 3$.

57. The GCF of $2a + 8$ is 2.
The GCF of $3a - 6$ is 3.
The GCF for the product is $2 \cdot 3 = 6$.

59. The GCF of $2x^2 + 5x$ is x.
The GCF of $7x - 14$ is 7.
The GCF for the product is $x \cdot 7 = 7x$.

61. $33t - t^2 = t \cdot 33 - t \cdot t = t(33 - t)$
The length of the rectangle is $33 - t$.

63. Writing exercise

65. Group exercise

Exercises 4.2

1. $x^2 - 8x + 15 = (x - 3)(x - 5)$

3. $m^2 + 8m + 12 = (m + 2)(m + 6)$

5. $p^2 - 8p - 20 = (p + 2)(p - 10)$

7. $x^2 - 16x + 64 = (x - 8)(x - 8)$

9. $x^2 - 7xy + 10y^2 = (x - 2y)(x - 5y)$

11. $x^2 + 8x + 15$
Consider only the positive factors of 15
because the last two terms are positive.
$15 = (1)(15)$
$ = (3)(5)$

Possible Factors	Middle Terms
$(x + 1)(x + 15)$	$16x$
$(x + 3)(x + 5)$	$8x$

$x^2 + 8x + 15 = (x + 3)(x + 5)$

13. $x^2 - 11x + 28$
Consider only the negative factors of 28
because the middle term is negative.
$28 = (-1)(-28)$
$ = (-2)(-14)$
$ = (-4)(-7)$

Possible Factors	Middle Terms
$(x - 1)(x - 28)$	$-29x$
$(x - 2)(x - 14)$	$-16x$
$(x - 4)(x - 7)$	$-11x$

$x^2 - 11x + 28 = (x - 4)(x - 7)$

15. $s^2 + 13s + 30$
Consider only the positive factors of 30
because the last two terms are positive.
$30 = (1)(30)$
$ = (2)(15)$
$ = (3)(10)$
$ = (5)(6)$

Possible Factors	Middle Terms
$(s + 1)(s + 30)$	$31s$
$(s + 2)(s + 15)$	$17s$
$(s + 3)(s + 10)$	$13s$
$(s + 5)(s + 6)$	$11s$

$s^2 + 13s + 30 = (s + 3)(s + 10)$

17. $a^2 - 2a - 48$
The factors of -48 will have different signs.
$-48 = (1)(-48)$
$ = (-1)(48)$
$ = (2)(-24)$
$ = (-2)(24)$
$ = (3)(-16)$
$ = (-3)(16)$
$ = (4)(-12)$
$ = (-4)(12)$
$ = (6)(-8)$
$ = (-6)(8)$

Possible Factors	Middle Terms
$(a + 1)(a - 48)$	$-47a$
$(a - 1)(a + 48)$	$47a$
$(a + 2)(a - 24)$	$-22a$
$(a - 2)(a + 24)$	$22a$
$(a + 3)(a - 16)$	$-13a$
$(a - 3)(a + 16)$	$13a$
$(a + 4)(a - 12)$	$-8a$
$(a - 4)(a + 12)$	$8a$
$(a + 6)(a - 8)$	$-2a$
$(a - 6)(a + 8)$	$2a$

$a^2 - 2a - 48 = (a + 6)(a - 8)$

19. $x^2 - 8x + 7$

Consider only the negative factors of 7 because the middle term is negative.

$7 = (-1)(-7)$

Possible Factors	Middle Terms
$(x - 1)(x - 7)$	$-8x$

$x^2 - 8x + 7 = (x - 1)(x - 7)$

21. $m^2 + 3m - 28$

The factors of -28 will have different signs.

$$-28 = (1)(-28)$$
$$= (-1)(28)$$
$$= (2)(-14)$$
$$= (-2)(14)$$
$$= (4)(-7)$$
$$= (-4)(7)$$

Possible Factors	Middle Terms
$(m + 1)(m - 28)$	$-27m$
$(m - 1)(m + 28)$	$27m$
$(m + 2)(m - 14)$	$-12m$
$(m - 2)(m + 14)$	$12m$
$(m + 4)(m - 7)$	$-3m$
$(m - 4)(m + 7)$	$3m$

$m^2 + 3m - 28 = (m - 4)(m + 7)$

23. $x^2 - 6x - 40$

The factors of -40 will have different signs.

$$-40 = (1)(-40)$$
$$= (-1)(40)$$
$$= (2)(-20)$$
$$= (-2)(20)$$
$$= (4)(-10)$$
$$= (-4)(10)$$
$$= (5)(-8)$$
$$= (-5)(8)$$

Possible Factors	Middle Terms
$(x + 1)(x - 40)$	$-39x$
$(x - 1)(x + 40)$	$39x$
$(x + 2)(x - 20)$	$-18x$
$(x - 2)(x + 20)$	$18x$
$(x + 4)(x - 10)$	$-6x$
$(x - 4)(x + 10)$	$6x$
$(x + 5)(x - 8)$	$-3x$
$(x - 5)(x + 8)$	$3x$

$x^2 - 6x - 40 = (x + 4)(x - 10)$

25. $x^2 - 14x + 49$

Consider only the negative factors of 49 because the middle term is negative.

$$49 = (-1)(-49)$$
$$= (-7)(-7)$$

Possible Factors	Middle Terms
$(x - 1)(x - 49)$	$-50x$
$(x - 7)(x - 7)$	$-14x$

$x^2 - 14x + 49 = (x - 7)(x - 7)$

27. $p^2 - 10p - 24$

The factors of -24 will have different signs.

$$-24 = (1)(-24)$$
$$= (-1)(24)$$
$$= (2)(-12)$$
$$= (-2)(12)$$
$$= (3)(-8)$$
$$= (-3)(8)$$
$$= (4)(-6)$$
$$= (-4)(6)$$

Possible Factors	Middle Terms
$(p + 1)(p - 24)$	$-23p$
$(p - 1)(p + 24)$	$23p$
$(p + 2)(p - 12)$	$-10p$
$(p - 2)(p + 12)$	$10p$
$(p + 3)(p - 8)$	$-5p$
$(p - 3)(p + 8)$	$5p$
$(p + 4)(p - 6)$	$-2p$
$(p - 4)(p + 6)$	$2p$

$$p^2 - 10p - 24 = (p + 2)(p - 12)$$

29. $x^2 + 5x - 66$

The factors of -66 will have different signs.

$$\begin{aligned}-66 &= (1)(-66) \\ &= (-1)(66) \\ &= (2)(-33) \\ &= (-2)(33) \\ &= (3)(-22) \\ &= (-3)(22) \\ &= (6)(-11) \\ &= (-6)(11)\end{aligned}$$

Possible Factors	Middle Terms
$(x + 1)(x - 66)$	$-65x$
$(x - 1)(x + 66)$	$65x$
$(x + 2)(x - 33)$	$-31x$
$(x - 2)(x + 33)$	$31x$
$(x + 3)(x - 22)$	$-19x$
$(x - 3)(x + 22)$	$19x$
$(x + 6)(x - 11)$	$-5x$
$(x - 6)(x + 11)$	$5x$

$$x^2 + 5x - 66 = (x - 6)(x + 11)$$

31. $c^2 + 19c + 60$

Consider only the positive factors of 60 because the last two terms are both positive.

$$\begin{aligned}60 &= (1)(60) \\ &= (2)(30) \\ &= (3)(20) \\ &= (4)(15) \\ &= (5)(12) \\ &= (6)(10)\end{aligned}$$

Possible Factors	Middle Terms
$(c + 1)(c + 60)$	$61c$
$(c + 2)(c + 30)$	$32c$
$(c + 3)(c + 20)$	$23c$
$(c + 4)(c + 15)$	$19c$

Stop when the right pair is found.
$$c^2 + 19c + 60 = (c + 4)(c + 15)$$

33. $n^2 + 5n - 50$

The factors of -50 will have different signs.

$$\begin{aligned}-50 &= (1)(-50) \\ &= (-1)(50) \\ &= (2)(-25) \\ &= (-2)(25) \\ &= (5)(-10) \\ &= (-5)(10)\end{aligned}$$

Possible Factors	Middle Terms
$(n + 1)(n - 50)$	$-49n$
$(n - 1)(n + 50)$	$49n$
$(n + 2)(n - 25)$	$-23n$
$(n - 2)(n + 25)$	$23n$
$(n + 5)(n - 10)$	$-5n$
$(n - 5)(n + 10)$	$5n$

$$n^2 + 5n - 50 = (n - 5)(n + 10)$$

35. $x^2 + 7xy + 10y^2$

Consider only the positive factors of $10y^2$ because the last two terms are positive.

$$\begin{aligned}10y^2 &= (y)(10y) \\ &= (2y)(5y)\end{aligned}$$

Possible Factors	Middle Terms
$(x + y)(x + 10y)$	$11xy$
$(x + 2y)(x + 5y)$	$7xy$

$$x^2 + 7xy + 10y^2 = (x + 2y)(x + 5y)$$

37. $a^2 - ab - 42b^2$

The factors of -42 will have different signs.

$$\begin{aligned}-42 &= (1)(-42) \\ &= (-1)(42) \\ &= (2)(-21) \\ &= (-2)(21) \\ &= (3)(-14) \\ &= (-3)(14) \\ &= (6)(-7) \\ &= (-6)(7)\end{aligned}$$

Possible Factors	Middle Terms
$(a + b)(a - 42b)$	$-41ab$
$(a - b)(a + 42b)$	$41ab$
$(a + 2b)(a - 21b)$	$-19ab$
$(a - 2b)(a + 21b)$	$19ab$
$(a + 3b)(a - 14b)$	$-11ab$
$(a - 3b)(a + 14b)$	$11ab$
$(a + 6b)(a - 7b)$	$-ab$

Stop when the right pair is found.
$$a^2 - ab - 42b^2 = (a + 6b)(a - 7b)$$

39. $x^2 - 13xy + 40y^2$

Consider only the negative factors of 40 because the middle term is negative.
$$\begin{aligned} 40 &= (-1)(-40) \\ &= (-2)(-20) \\ &= (-4)(-10) \\ &= (-5)(-8) \end{aligned}$$

Possible Factors	Middle Terms
$(x - y)(x - 40y)$	$-41xy$
$(x - 2y)(x - 20y)$	$-22xy$
$(x - 4y)(x - 10y)$	$-14xy$
$(x - 5y)(x - 8y)$	$-13xy$

$$x^2 - 13xy + 40y^2 = (x - 5y)(x - 8y)$$

41. $b^2 + 6ab + 9a^2$

Consider only the positive factors of 9 because the last two terms are positive.
$$\begin{aligned} 9 &= (1)(9) \\ &= (3)(3) \end{aligned}$$

Possible Factors	Middle Terms
$(b + a)(b + 9a)$	$10ab$
$(b + 3a)(b + 3a)$	$6ab$

$$b^2 + 6ab + 9a^2 = (b + 3a)(b + 3a)$$

43. $x^2 - 2xy - 8y^2$

The factors of -8 will have different signs.
$$\begin{aligned} -8 &= (1)(-8) \\ &= (-1)(8) \\ &= (2)(-4) \\ &= (-2)(4) \end{aligned}$$

Possible Factors	Middle Terms
$(x + y)(x - 8y)$	$-7xy$
$(x - y)(x + 8y)$	$7xy$
$(x + 2y)(x - 4y)$	$-2xy$
$(x - 2y)(x + 4y)$	$2xy$

$$x^2 - 2xy - 8y^2 = (x + 2y)(x - 4y)$$

45. $25m^2 + 10mn + n^2 = n^2 + 10mn + 25m^2$

Consider only the positive factors of 25 because the last two terms are positive.
$$\begin{aligned} 25 &= (1)(25) \\ &= (5)(5) \end{aligned}$$

Possible Factors	Middle Terms
$(n + m)(n + 25m)$	$26mn$
$(n + 5m)(n + 5m)$	$10mn$

$$25m^2 + 10mn + n^2 = (n + 5m)(n + 5m)$$

47. $\begin{aligned} 3a^2 - 3a - 126 &= 3(a^2 - a - 42) \\ &= 3(a + 6)(a - 7) \end{aligned}$

49. $\begin{aligned} r^3 + 7r^2 - 18r &= r(r^2 + 7r - 18) \\ &= r(r - 2)(r + 9) \end{aligned}$

51. $\begin{aligned} 2x^3 - 20x^2 - 48x &= 2x(x^2 - 10x - 24) \\ &= 2x(x - 12)(x + 2) \end{aligned}$

53. $\begin{aligned} x^2y - 9xy^2 - 36y^3 &= y(x^2 - 9xy - 36y^2) \\ &= y(x + 3y)(x - 12y) \end{aligned}$

55. $\begin{aligned} m^3 - 29m^2n + 120mn^2 &= m(m^2 - 29mn + 120n^2) \\ &= m(m - 5n)(m - 24n) \end{aligned}$

57. $x^2 + kx + 8$

Consider the positive factors of 8.
$$\begin{aligned} 8 &= (1)(8) \\ &= (2)(4) \end{aligned}$$

Possible Factors	Middle Terms
$(x + 1)(x + 8)$	$9x$
$(x + 2)(x + 4)$	$6x$

$k = 6$ or 9

59. $x^2 - kx + 16$

Consider the negative factors of 16.

$$16 = (-1)(-16)$$
$$= (-2)(-8)$$
$$= (-4)(-4)$$

Possible Factors	Middle Terms
$(x - 1)(x - 16)$	$-17x$
$(x - 2)(x - 8)$	$-10x$
$(x - 4)(x - 4)$	$-8x$

$k = 8$, 10, or 17

61. $x^2 - kx - 5$

The factors of -5 will have different signs.

$$-5 = (1)(-5)$$
$$= (-1)(5)$$

Possible Factors	Middle Terms
$(x + 1)(x - 5)$	$-4x$
$(x - 1)(x + 5)$	$4x$

$k = 4$

63. $x^2 + 3x + k$

Since the sign of the last two terms is positive, then we need to find all pairs of positive integers whose sum is 3. $1 + 2 = 3$, and since $2 = (1)(2)$ and $(x + 1)(x + 2) = x^2 + 3x + 2$, then $k = 2$.

65. $x^2 + 2x - k$

Since the last term is negative, then we need to find pairs of integers of opposite sign whose sum is 2. $-1 + 3 = 2$, $-2 + 4 = 2$, $-3 + 5 = 2$, etc. So, since

$$-3 = (-1)(3) \text{ and } (x - 1)(x + 3) = x^2 + 2x - 3$$
$$-8 = (-2)(4) \text{ and } (x - 2)(x + 4) = x^2 + 2x - 8$$
$$-15 = (-3)(5) \text{ and } (x - 3)(x + 5) = x^2 + 2x - 15$$
$$-24 = (-4)(6) \text{ and } (x - 4)(x + 6) = x^2 + 2x - 24$$
     etc.,

then $k = 3$, 8, 15, 24, …

Exercises 4.3

1. $4x^2 - 4x - 3 = (2x + 1)(2x - 3)$

3. $6a^2 + 13a + 6 = (2a + 3)(3a + 2)$

5. $15x^2 - 16x + 4 = (3x - 2)(5x - 2)$

7. $16a^2 + 8ab + b^2 = (4a + b)(4a + b)$

9. $4m^2 + 5mn - 6n^2 = (m + 2n)(4m - 3n)$

11. $3x^2 + 7x + 2$

$3 = 3 \cdot 1 \qquad 2 = 2 \cdot 1$

Notice all the signs of the trinomial are positive.

Possible Factors	Middle Terms
$(3x + 2)(x + 1)$	$5x$
$(3x + 1)(x + 2)$	$7x$

$3x^2 + 7x + 2 = (3x + 1)(x + 2)$

13. $2w^2 + 13w + 15$

$2 = 2 \cdot 1 \qquad 15 = 15 \cdot 1$
$\qquad\qquad\qquad = 5 \cdot 3$

Notice all the signs of the trinomial are positive.

Possible Factors	Middle Terms
$(2w + 15)(w + 1)$	$17w$
$(2w + 1)(w + 15)$	$31w$
$(2w + 5)(w + 3)$	$11w$
$(2w + 3)(w + 5)$	$13w$

$2w^2 + 13w + 15 = (2w + 3)(w + 5)$

15. $5x^2 - 16x + 3$

$5 = 5 \cdot 1 \qquad 3 = 3 \cdot 1$

Notice only the middle term is negative.

Possible Factors	Middle Terms
$(5x - 3)(x - 1)$	$-8x$
$(5x - 1)(x - 3)$	$-16x$

$5x^2 - 16x + 3 = (5x - 1)(x - 3)$

17. $4x^2 - 12x + 5$

$\quad 4 = 4 \cdot 1 \qquad 5 = 5 \cdot 1$

$\quad\quad = 2 \cdot 2$

Notice only the middle term is negative.

Possible Factors	Middle Terms
$(4x - 5)(x - 1)$	$-9x$
$(4x - 1)(x - 5)$	$-21x$
$(2x - 5)(2x - 1)$	$-12x$

$4x^2 - 12x + 5 = (2x - 5)(2x - 1)$

19. $3x^2 - 5x - 2$

$\quad 3 = 3 \cdot 1 \qquad 2 = 2 \cdot 1$

Notice two of the signs are negative.

Possible Factors	Middle Terms
$(3x + 2)(x - 1)$	$-x$
$(3x - 2)(x + 1)$	x
$(3x + 1)(x - 2)$	$-5x$
$(3x - 1)(x + 2)$	$5x$

$3x^2 - 5x - 2 = (3x + 1)(x - 2)$

21. $4p^2 + 19p - 5$

$\quad 4 = 4 \cdot 1 \qquad 5 = 5 \cdot 1$

$\quad\quad = 2 \cdot 2$

Notice only the last term is negative.

Possible Factors	Middle Terms
$(4p + 5)(p - 1)$	p
$(4p - 5)(p + 1)$	$-p$
$(4p + 1)(p - 5)$	$-19p$
$(4p - 1)(p + 5)$	$19p$

Stop as soon as the correct pair is found.

$4p^2 + 19p - 5 = (4p - 1)(p + 5)$

23. $6x^2 + 19x + 10$

$\quad 6 = 6 \cdot 1 \qquad 10 = 10 \cdot 1$

$\quad\quad = 3 \cdot 2 \qquad\quad = 5 \cdot 2$

Notice all the signs of the trinomial are positive.

Possible Factors	Middle Terms
$(6x + 10)(x + 1)$	$16x$
$(6x + 1)(x + 10)$	$61x$
$(6x + 5)(x + 2)$	$17x$
$(6x + 2)(x + 5)$	$32x$
$(3x + 10)(2x + 1)$	$23x$
$(3x + 1)(2x + 10)$	$32x$
$(3x + 5)(2x + 2)$	$16x$
$(3x + 2)(2x + 5)$	$19x$

$6x^2 + 19x + 10 = (3x + 2)(2x + 5)$

25. $15x^2 + x - 6$

$\quad 15 = 15 \cdot 1 \qquad 6 = 6 \cdot 1$

$\quad\quad = 5 \cdot 3 \qquad\quad = 3 \cdot 2$

Notice only the last term is negative.

Possible Factors	Middle Terms
$(15x + 6)(x - 1)$	$-9x$
$(15x - 6)(x + 1)$	$9x$
$(15x + 1)(x - 6)$	$-89x$
$(15x - 1)(x + 6)$	$89x$
$(5x + 3)(3x - 2)$	$-x$
$(5x - 3)(3x + 2)$	x

Stop as soon as the correct pair is found.

$15x^2 + x - 6 = (5x - 3)(3x + 2)$

27. $6m^2 + 25m - 25$

$\quad 6 = 6 \cdot 1 \qquad 25 = 25 \cdot 1$

$\quad\quad = 3 \cdot 2 \qquad\quad = 5 \cdot 5$

Notice only the last term is negative.

Possible Factors	Middle Terms
$(6m + 25)(m - 1)$	$19m$
$(6m - 25)(m + 1)$	$-19m$
$(6m + 5)(m - 5)$	$-25m$
$(6m - 5)(m + 5)$	$25m$

Stop as soon as the correct pair is found.
$6m^2 + 25m - 25 = (6m - 5)(m + 5)$

29. $9x^2 - 12x + 4$
$9 = 3 \cdot 3 \qquad 4 = 2 \cdot 2$
$ = 9 \cdot 1 \qquad = 4 \cdot 1$
Consider only the negative factors of 4.

Possible Factors	Middle Terms
$(3x - 2)(3x - 2)$	$-12x$

Stop as soon as the correct pair is found.
$9x^2 - 12x + 4 = (3x - 2)(3x - 2)$

31. $12x^2 - 8x - 15$
$12 = 4 \cdot 3 \qquad 15 = 5 \cdot 3$
$ = 6 \cdot 2 \qquad = 15 \cdot 1$
$ = 12 \cdot 1$
Notice there are two negative signs.

Possible Factors	Middle Terms
$(4x - 5)(3x + 3)$	$-3x$
$(4x + 5)(3x - 3)$	$3x$
$(4x - 3)(3x + 5)$	$11x$
$(4x + 3)(3x - 5)$	$-11x$
$(6x - 5)(2x + 3)$	$8x$
$(6x + 5)(2x - 3)$	$-8x$

Stop as soon as the correct pair is found.
$12x^2 - 8x - 15 = (6x + 5)(2x - 3)$

33. $3y^2 + 7y - 6$
$3 = 3 \cdot 1 \qquad 6 = 6 \cdot 1$
$ = 3 \cdot 2$
Notice only one sign is negative.

Possible Factors	Middle Terms
$(3y + 6)(y - 1)$	$3y$
$(3y - 6)(y + 1)$	$-3y$
$(3y + 3)(y - 2)$	$-3y$
$(3y - 3)(y + 2)$	$3y$
$(3y + 2)(y - 3)$	$-7y$
$(3y - 2)(y + 3)$	$7y$

Stop as soon as the correct pair is found.
$3y^2 + 7y - 6 = (3y - 2)(y + 3)$

35. $8x^2 - 27x - 20$
$8 = 8 \cdot 1 \qquad 20 = 20 \cdot 1$
$ = 4 \cdot 2 \qquad = 5 \cdot 4$
$ = 10 \cdot 2$
Notice two of the signs are negative.

Possible Factors	Middle Terms
$(8x + 20)(x - 1)$	$12x$
$(8x - 20)(x + 1)$	$-12x$
$(8x + 1)(x - 20)$	$-159x$
$(8x - 1)(x + 20)$	$159x$
$(8x + 5)(x - 4)$	$-27x$

Stop as soon as the correct pair is found.
$8x^2 - 27x - 20 = (8x + 5)(x - 4)$

37. $2x^2 + 3xy + y^2$
$2 = 2 \cdot 1 \qquad 1 = 1 \cdot 1$
Notice all of the signs of the trinomial are positive.

Possible Factors	Middle Terms
$(2x + y)(x + y)$	$3xy$

$2x^2 + 3xy + y^2 = (2xy + y)(x + y)$

39. $5a^2 - 8ab - 4b^2$
$5 = 5 \cdot 1 \qquad 4 = 4 \cdot 1$
$ = 2 \cdot 2$
Notice two of the signs are negative.

Possible Factors	Middle Terms
$(5a + 4b)(a - b)$	$-ab$
$(5a - 4b)(a + b)$	ab
$(5a + b)(a - 4b)$	$-19ab$
$(5a - b)(a + 4b)$	$19ab$
$(5a + 2b)(a - 2b)$	$-8ab$

$5a^2 - 8ab - 4b^2 = (5a + 2b)(a - 2b)$

41. $9x^2 + 4xy - 5y^2$
$9 = 9 \cdot 1 \qquad 5 = 5 \cdot 1$
$ = 3 \cdot 3$
Notice only one of the signs is negative.

Possible Factors	Middle Terms
$(9x + 5y)(x - y)$	$-4xy$
$(9x - 5y)(x + y)$	$4xy$

Stop as soon as the correct pair is found.
$$9x^2 + 4xy - 5y^2 = (9x - 5y)(x + y)$$

43. $6m^2 - 17mn + 12n^2$

$$6 = 6 \cdot 1 \qquad 12 = 12 \cdot 1$$
$$= 3 \cdot 2 \qquad = 4 \cdot 3$$

Consider only the negative factors of 12.

Possible Factors	Middle Terms
$(6m - 12n)(m - n)$	$-18mn$
$(6m - n)(m - 12n)$	$-73mn$
$(6m - 4n)(m - 3n)$	$-22mn$
$(6m - 3n)(m - 4n)$	$-27mn$
$(3m - 4n)(2m - 3n)$	$-17mn$

Stop as soon as the correct pair is found.
$$6m^2 - 17mn + 12n^2 = (3m - 4n)(2m - 3n)$$

45. $36a^2 - 3ab - 5b^2$

$$36 = 36 \cdot 1 \qquad 5 = 5 \cdot 1$$
$$= 18 \cdot 2$$
$$= 12 \cdot 3$$
$$= 9 \cdot 4$$
$$= 6 \cdot 6$$

Notice two of the signs are negative.

Possible Factors	Middle Terms
$(36a + 5b)(a - b)$	$-31ab$
$(36a - 5b)(a + b)$	$31ab$
$(36a + b)(a - 5b)$	$-179ab$
$(36a - b)(a + 5b)$	$179ab$
$(18a + 5b)(2a - b)$	$-8ab$
$(18a - 5b)(2a + b)$	$8ab$
$(18a + b)(2a - 5b)$	$-88ab$
$(18a - b)(2a + 5b)$	$88ab$
$(12a + 5b)(3a - b)$	$3ab$
$(12a - 5b)(3a + b)$	$-3ab$

Stop as soon as the correct pair is found.
$$36a^2 - 3ab - 5b^2 = (12a - 5b)(3a + b)$$

47. $x^2 + 4xy + 4y^2$

$$1 = 1 \cdot 1 \qquad 4 = 4 \cdot 1$$
$$= 2 \cdot 2$$

Notice all of the signs are positive.

Possible Factors	Middle Terms
$(x + 4y)(x + y)$	$5xy$
$(x + 2y)(x + 2y)$	$4xy$

$$x^2 + 4xy + 4y^2 = (x + 2y)(x + 2y)$$

49. $20x^2 - 20x - 15 = 5(4x^2 - 4x - 3)$
$$= 5(2x - 3)(2x + 1)$$

51. $8m^2 + 12m + 4 = 4(2m^2 + 3m + 1)$
$$= 4(2m + 1)(m + 1)$$

53. $15r^2 - 21rs + 6s^2 = 3(5r^2 - 7rs + 2s^2)$
$$= 3(5r - 2s)(r - s)$$

55. $2x^3 - 2x^2 - 4x = 2x(x^2 - x - 2)$
$$= 2x(x - 2)(x + 1)$$

57. $2y^4 + 5y^3 + 3y^2 = y^2(2y^2 + 5y + 3)$
$$= y^2(2y + 3)(y + 1)$$

59. $36a^3 - 66a^2 + 18a = 6a(6a^2 - 11a + 3)$
$$= 6a(3a - 1)(2a - 3)$$

61. $9p^2 + 30pq + 21q^2 = 3(3p^2 + 10pq + 7q^2)$
$$= 3(p + q)(3p + 7q)$$

63. $10(x + y)^2 - 11(x + y) - 6$
$$= [5(x + y) + 2][2(x + y) - 3]$$
$$= (5x + 5y + 2)(2x + 2y - 3)$$

65. $5(x - 1)^2 - 15(x - 1) - 350$
$$= 5[(x - 1)^2 - 3(x - 1) - 70]$$
$$= 5[(x - 1) - 10][(x - 1) + 7]$$
$$= 5(x - 1 - 10)(x - 1 + 7)$$
$$= 5(x - 11)(x + 6)$$

67. $15 + 29x - 48x^2 = (1 + 3x)(15 - 16x)$

69. $-6x^2 + 19x - 15 = (3x - 5)(-2x + 3)$

Exercises 4.4

1. The binomial $3x^2 + 2y^2$ is not the difference of squares.

3. Since $16a^2 - 25b^2 = (4a)^2 - (5b)^2$, it is the difference of squares.

5. The binomial $16r^2 + 4$ is not the difference of squares.

7. The binomial $16a^2 - 12b^2$ is not the difference of squares.

9. Since $a^2b^2 - 25 = (ab)^2 - (5)^2$, it is the difference of squares.

11. $m^2 - n^2 = (m+n)(m-n)$

13. $x^2 - 49 = (x+7)(x-7)$

15. $49 - y^2 = (7+y)(7-y)$

17. $9b^2 - 16 = (3b+4)(3b-4)$

19. $16w^2 - 49 = (4w+7)(4w-7)$

21. $4s^2 - 9r^2 = (2s+3r)(2s-3r)$

23. $9w^2 - 49z^2 = (3w+7z)(3w-7z)$

25. $16a^2 - 49b^2 = (4a+7b)(4a-7b)$

27. $x^4 - 36 = (x^2+6)(x^2-6)$

29. $x^2y^2 - 16 = (xy+4)(xy-4)$

31. $25 - a^2b^2 = (5+ab)(5-ab)$

33. $r^4 - 4s^2 = (r^2+2s)(r^2-2s)$

35. $81a^2 - 100b^6 = (9a+10b^3)(9a-10b^3)$

37. $18x^3 - 2xy^2 = 2x(9x^2 - y^2)$
$\qquad = 2x(3x+y)(3x-y)$

39. $12m^3n - 75mn^3 = 3mn(4m^2 - 25n^2)$
$\qquad = 3mn(2m+5n)(2m-5n)$

41. $48a^2b^2 - 27b^4 = 3b^2(16a^2 - 9b^2)$
$\qquad = 3b^2(4a+3b)(4a-3b)$

43. $x^2 - 14x + 49$ is a perfect square trinomial in which $a = x$ and $b = -7$.
$x^2 - 14x + 49 = (x-7)^2$

45. $x^2 - 18x - 81$ is not a perfect square trinomial.

47. $x^2 - 18x + 81$ is a perfect square trinomial in which $a = x$ and $b = -9$.
$x^2 - 18x + 81 = (x-9)^2$

49. $a = x, b = 2$
$x^2 + 4x + 4 = (x+2)^2$

51. $a = x, b = -5$
$x^2 - 10x + 25 = (x-5)^2$

53. $a = 2x, b = 3y$
$4x^2 + 12xy + 9y^2 = (2x+3y)^2$

55. $a = 3x, b = -4y$
$9x^2 - 24xy + 16y^2 = (3x-4y)^2$

57. $y^3 - 10y^2 + 25y = y(y^2 - 10y + 25) = y(y-5)^2$
$a = y, b = -5$

59. $x^2(x+y) - y^2(x+y) = (x+y)(x^2 - y^2)$
$\qquad = (x+y)(x+y)(x-y)$
$\qquad = (x+y)^2(x-y)$

61. $2m^2(m-2n) - 18n^2(m-2n)$
$\qquad = 2(m-2n)(m^2 - 9n^2)$
$\qquad = 2(m-2n)(m+3n)(m-3n)$

63. $kx^2 - 25 = (2x+5)(2x-5)$
$kx^2 - 25 = 4x^2 - 25$ so $k = 4$

65. $2x^3 - kxy^2 = 2x(x-3y)(x+3y)$
$\qquad = 2x(x^2 - 9y^2)$
$\qquad = 2x^3 - 18xy^2$ so $k = 18$

67. Writing exercise

Exercises 4.5

1. $x^3 - 4x^2 + 3x - 12 = (x^3 - 4x^2) + (3x - 12)$
$$= x^2(x - 4) + 3(x - 4)$$
$$= (x - 4)(x^2 + 3)$$

3. $a^3 - 3a^2 + 5a - 15 = (a^3 - 3a^2) + (5a - 15)$
$$= a^2(a - 3) + 5(a - 3)$$
$$= (a - 3)(a^2 + 5)$$

5. $10x^3 + 5x^2 - 2x - 1 = (10x^3 + 5x^2) + (-2x - 1)$
$$= 5x^2(2x + 1) - 1(2x + 1)$$
$$= (2x + 1)(5x^2 - 1)$$

7. $x^4 - 2x^3 + 3x - 6 = (x^4 - 2x^3) + (3x - 6)$
$$= x^3(x - 2) + 3(x - 2)$$
$$= (x - 2)(x^3 + 3)$$

9. $3x - 6 + xy - 2y = (3x - 6) + (xy - 2y)$
$$= 3(x - 2) + y(x - 2)$$
$$= (x - 2)(3 + y)$$

11. $ab - ac + b^2 - bc = (ab - ac) + (b^2 - bc)$
$$= a(b - c) + b(b - c)$$
$$= (b - c)(a + b)$$

13. $3x^2 - 2xy + 3x - 2y = (3x^2 - 2xy) + (3x - 2y)$
$$= x(3x - 2y) + 1(3x - 2y)$$
$$= (3x - 2y)(x + 1)$$

15. $5s^2 + 15st - 2st - 6t^2 = (5s^2 + 15st) + (-2st - 6t^2)$
$$= 5s(s + 3t) - 2t(s + 3t)$$
$$= (s + 3t)(5s - 2t)$$

17. $3x^3 + 6x^2y - x^2y - 2xy^2$
$$= x(3x^2 + 6xy - xy - 2y^2)$$
$$= x[(3x^2 + 6xy) + (-xy - 2y^2)]$$
$$= x[3x(x + 2y) - y(x + 2y)]$$
$$= x(x + 2y)(3x - y)$$

19. $x^4 + 5x^3 - 2x^2 - 10x$
$$= x(x^3 + 5x^2 - 2x - 10)$$
$$= x[(x^3 + 5x^2) + (-2x - 10)]$$
$$= x[x^2(x + 5) - 2(x + 5)]$$
$$= x(x + 5)(x^2 - 2)$$

21. $2x^3 - 2x^2 + 3x^2 - 3x$
$$= x(2x^2 - 2x + 3x - 3)$$
$$= x[(2x^2 - 2x) + (3x - 3)]$$
$$= x[2x(x - 1) + 3(x - 1)]$$
$$= x(x - 1)(2x + 3)$$

Exercises 4.6

1. $x^2 + 2x - 3 \stackrel{?}{=} (x + 3)(x - 1)$; True
$$(x + 3)(x - 1) = x^2 - x + 3x - 3$$
$$= x^2 + 2x - 3$$

3. $x^2 - 10x - 24 \stackrel{?}{=} (x - 6)(x + 4)$; False
$$(x - 6)(x + 4) = x^2 + 4x - 6x - 24$$
$$= x^2 - 2x - 24$$

5. $x^2 - 16x + 64 \stackrel{?}{=} (x - 8)(x - 8)$; True
$$(x - 8)(x - 8) = x^2 - 8x - 8x + 64$$
$$= x^2 - 16x + 64$$

7. $25y^2 - 10y + 1 \stackrel{?}{=} (5y - 1)(5y + 1)$; False
$$(5y - 1)(5y + 1) = (5y)^2 - 1^2$$
$$= 25y^2 - 1$$

9. $10p^2 - pq - 3q^2 \stackrel{?}{=} (5p - 3q)(2p + q)$; True
$$(5p - 3q)(2p + q) = 10p^2 + 5qp - 6pq - 3q^2$$
$$= 10p^2 - pq - 3q^2$$

11. $x^2 + 4x - 9$; $a = 1$, $b = 4$, $c = -9$

13. $x^2 - 3x + 8$; $a = 1$, $b = -3$, $c = 8$

15. $3x^2 + 5x - 8$; $a = 3$, $b = 5$, $c = -8$

17. $4x^2 + 8x + 11$; $a = 4$, $b = 8$, $c = 11$

19. $-3x^2 + 5x - 10$; $a = -3$, $b = 5$, $c = -10$

21. $x^2 - x - 6$
$a = 1 \quad b = -1 \quad c = -6$
$mn = ac = 1(-6) = -6$
$m + n = b = -1$

mn	$m + n$
$1(-6) = -6$	$1 + (-6) = -5$
$2(-3) = -6$	$2 + (-3) = -1$

We need look no further than 2 and –3.
Because we found values for m and n, the trinomial is factorable.
m and n are 2 and –3.

23. $x^2 + x + 2$
$a = 1 \quad b = 1 \quad c = 2$
$mn = ac = 1(2) = 2$
$m + n = b = 1$

mn	$m + n$
$1(2) = 2$	$1 + 2 = 3$
$-1(-2) = 2$	$-1 + (-2) = -3$

None of these pairs has a sum of 1.
The trinomial is not factorable.

25. $x^2 - 5x + 6$
$a = 1 \quad b = -5 \quad c = 6$
$mn = ac = 1(6) = 6$
$m + n = b = -5$

mn	$m + n$
$1(6) = 6$	$1 + 6 = 7$
$2(3) = 6$	$2 + 3 = 5$
$-1(-6) = 6$	$-1 + (-6) = -7$
$-2(-3) = 6$	$-2 + (-3) = -5$

Because we found values for m and n, the trinomial is factorable.
m and n are –3 and –2.

27. $2x^2 + 5x - 3$
$a = 2 \quad b = 5 \quad c = -3$
$mn = ac = 2(-3) = -6$
$m + n = b = 5$

mn	$m + n$
$1(-6) = -6$	$1 + (-6) = -5$
$2(-3) = -6$	$2 + (-3) = -1$
$3(-2) = -6$	$3 + (-2) = 1$
$6(-1) = -6$	$6 + (-1) = 5$

Because we found values for m and n, the trinomial is factorable.
m and n are 6 and –1.

29. $6x^2 - 19x + 10$
$a = 6 \quad b = -19 \quad c = 10$
$mn = ac = 6(10) = 60$
$m + n = b = -19$

mn	$m + n$
$-1(-60) = 60$	$-1 + (-60) = -61$
$-2(-30) = 60$	$-2 + (-30) = -32$
$-3(-20) = 60$	$-3 + (-20) = -23$
$-4(-15) = 60$	$-4 + (-15) = -19$

We need look no further than –4 and –15.
Because we found values for m and n, the trinomial is factorable.

31. $x^2 + 6x + 8 = x^2 + 4x + 2x + 8$
$\quad = (x^2 + 4x) + (2x + 8)$
$\quad = x(x + 4) + 2(x + 4)$
$\quad = (x + 2)(x + 4)$

33. $x^2 - 9x + 20 = x^2 - 5x - 4x + 20$
$\quad = (x^2 - 5x) + (-4x + 20)$
$\quad = x(x - 5) - 4(x - 5)$
$\quad = (x - 5)(x - 4)$

35. $x^2 - 2x - 63 = x^2 - 9x + 7x - 63$
$\quad = (x^2 - 9x) + (7x - 63)$
$\quad = x(x - 9) + 7(x - 9)$
$\quad = (x - 9)(x + 7)$

37. $x^2 + 8x + 15 = x^2 + 5x + 3x + 15$
$\quad = (x^2 + 5x) + (3x + 15)$
$\quad = x(x + 5) + 3(x + 5)$
$\quad = (x + 3)(x + 5)$

39. $x^2 - 11x + 28 = x^2 - 7x - 4x + 28$
$\quad = (x^2 - 7x) + (-4x + 28)$
$\quad = x(x - 7) - 4(x - 7)$
$\quad = (x - 4)(x - 7)$

41. $s^2 + 13s + 30 = s^2 + 3s + 10s + 30$
$\quad = (s^2 + 3s) + (10s + 30)$
$\quad = s(s + 3) + 10(s + 3)$
$\quad = (s + 10)(s + 3)$

43. $a^2 - 2a - 48 = a^2 - 8a + 6a - 48$
$$= (a^2 - 8a) + (6a - 48)$$
$$= a(a - 8) + 6(a - 8)$$
$$= (a - 8)(a + 6)$$

45. $x^2 - 8x + 7 = x^2 - 7x - x + 7$
$$= (x^2 - 7x) + (-x + 7)$$
$$= x(x - 7) - (x - 7)$$
$$= (x - 1)(x - 7)$$

47. $x^2 - 6x - 40 = x^2 - 10x + 4x - 40$
$$= (x^2 - 10x) + (4x - 40)$$
$$= x(x - 10) + 4(x - 10)$$
$$= (x - 10)(x + 4)$$

49. $x^2 - 14x + 49 = x^2 - 7x - 7x + 49$
$$= (x^2 - 7x) + (-7x + 49)$$
$$= x(x - 7) - 7(x - 7)$$
$$= (x - 7)(x - 7)$$

51. $p^2 - 10p - 24 = p^2 - 12p + 2p - 24$
$$= (p^2 - 12p) + (2p - 24)$$
$$= p(p - 12) + 2(p - 12)$$
$$= (p - 12)(p + 2)$$

53. $x^2 + 5x - 66 = x^2 + 11x - 6x - 66$
$$= (x^2 + 11x) + (-6x - 66)$$
$$= x(x + 11) - 6(x + 11)$$
$$= (x + 11)(x - 6)$$

55. $c^2 + 19c + 60 = c^2 + 15c + 4c + 60$
$$= (c^2 + 15c) + (4c + 60)$$
$$= c(c + 15) + 4(c + 15)$$
$$= (c + 4)(c + 15)$$

57. $n^2 + 5n - 50 = n^2 + 10n - 5n - 50$
$$= (n^2 + 10n) + (-5n - 50)$$
$$= n(n + 10) - 5(n + 10)$$
$$= (n + 10)(n - 5)$$

59. $x^2 + 7xy + 10y^2 = x^2 + 5xy + 2xy + 10y^2$
$$= (x^2 + 5xy) + (2xy + 10y^2)$$
$$= x(x + 5y) + 2y(x + 5y)$$
$$= (x + 2y)(x + 5y)$$

61. $a^2 - ab - 42b^2 = a^2 - 7ab + 6ab - 42b^2$
$$= (a^2 - 7ab) + (6ab - 42b^2)$$
$$= a(a - 7b) + 6b(a - 7b)$$
$$= (a - 7b)(a + 6b)$$

63. $x^2 - 13xy + 40y^2 = x^2 - 8xy - 5xy + 40y^2$
$$= (x^2 - 8xy) + (-5xy + 40y^2)$$
$$= x(x - 8y) - 5y(x - 8y)$$
$$= (x - 5y)(x - 8y)$$

65. $6x^2 + 19x + 10 = 6x^2 + 4x + 15x + 10$
$$= (6x^2 + 4x) + (15x + 10)$$
$$= 2x(3x + 2) + 5(3x + 2)$$
$$= (3x + 2)(2x + 5)$$

67. $15x^2 + x - 6 = 15x^2 - 9x + 10x - 6$
$$= (15x^2 - 9x) + (10x - 6)$$
$$= 3x(5x - 3) + 2(5x - 3)$$
$$= (5x - 3)(3x + 2)$$

69. $6m^2 + 25m - 25 = 6m^2 - 5m + 30m - 25$
$$= (6m^2 - 5m) + (30m - 25)$$
$$= m(6m - 5) + 5(6m - 5)$$
$$= (6m - 5)(m + 5)$$

71. $9x^2 - 12x + 4 = 9x^2 - 6x - 6x + 4$
$$= (9x^2 - 6x) + (-6x + 4)$$
$$= 3x(3x - 2) - 2(3x - 2)$$
$$= (3x - 2)(3x - 2)$$

73. $12x^2 - 8x - 15 = 12x^2 + 10x - 18x - 15$
$$= (12x^2 + 10x) + (-18x - 15)$$
$$= 2x(6x + 5) - 3(6x + 5)$$
$$= (6x + 5)(2x - 3)$$

75. $3y^2 + 7y - 6 = 3y^2 - 2y + 9y - 6$
$$= (3y^2 - 2y) + (9y - 6)$$
$$= y(3y - 2) + 3(3y - 2)$$
$$= (3y - 2)(y + 3)$$

77. $8x^2 - 27x - 20 = 8x^2 + 5x - 32x - 20$
$$= (8x^2 + 5x) + (-32x - 20)$$
$$= x(8x + 5) - 4(8x + 5)$$
$$= (8x + 5)(x - 4)$$

79. $2x^2 + 3xy + y^2 = 2x^2 + xy + 2xy + y^2$
$$= (2x^2 + xy) + (2xy + y^2)$$
$$= x(2x + y) + y(2x + y)$$
$$= (2x + y)(x + y)$$

81. $5a^2 - 8ab - 4b^2 = 5a^2 + 2ab - 10ab - 4b^2$
$$= (5a^2 + 2ab) + (-10ab - 4b^2)$$
$$= a(5a + 2b) - 2b(5a + 2b)$$
$$= (5a + 2b)(a - 2b)$$

83. $9x^2 + 4xy - 5y^2 = 9x^2 - 5xy + 9xy - 5y^2$
$$= (9x^2 - 5xy) + (9xy - 5y^2)$$
$$= x(9x - 5y) + y(9x - 5y)$$
$$= (9x - 5y)(x + y)$$

85. $6m^2 - 17mn + 12n^2$
$$= 6m^2 - 8mn - 9mn + 12n^2$$
$$= (6m^2 - 8mn) + (-9mn + 12n^2)$$
$$= 2m(3m - 4n) - 3n(3m - 4n)$$
$$= (3m - 4n)(2m - 3n)$$

87. $36a^2 - 3ab - 5b^2 = 36a^2 - 15ab + 12ab - 5b^2$
$$= (36a^2 - 15ab) + (12ab - 5b^2)$$
$$= 3a(12a - 5b) + b(12a - 5b)$$
$$= (12a - 5b)(3a + b)$$

89. $x^2 + 4xy + 4y^2 = x^2 + 2xy + 2xy + 4y^2$
$$= (x^2 + 2xy) + (2xy + 4y^2)$$
$$= x(x + 2y) + 2y(x + 2y)$$
$$= (x + 2y)(x + 2y)$$
$$= (x + 2y)^2$$

91. $20x^2 - 20x - 15 = 5(4x^2 - 4x - 3)$
$$= 5(4x^2 - 6x + 2x - 3)$$
$$= 5[2x(2x - 3) + 1(2x - 3)]$$
$$= 5(2x - 3)(2x + 1)$$

93. $8m^2 + 12m + 4 = 4(2m^2 + 3m + 1)$
$$= 4(2m^2 + m + 2m + 1)$$
$$= 4[m(2m + 1) + (2m + 1)]$$
$$= 4(2m + 1)(m + 1)$$

95. $15r^2 - 21rs + 6s^2 = 3(5r^2 - 7rs + 2s^2)$
$$= 3(5r^2 - 2rs - 5rs + 2s^2)$$
$$= 3[r(5r - 2s) - s(5r - 2s)]$$
$$= 3(5r - 2s)(r - s)$$

97. $2x^3 - 2x^2 - 4x = 2x(x^2 - x - 2)$
$$= 2x(x^2 - 2x + x - 2)$$
$$= 2x[x(x - 2) + (x - 2)]$$
$$= 2x(x - 2)(x + 1)$$

99. $2y^4 + 5y^3 + 3y^2 = y^2(2y^2 + 5y + 3)$
$$= y^2(2y^2 + 3y + 2y + 3)$$
$$= y^2[y(2y + 3) + (2y + 3)]$$
$$= y^2(2y + 3)(y + 1)$$

101. $36a^3 - 66a^2 + 18a = 6a(6a^2 - 11a + 3)$
$$= 6a(6a^2 - 2a - 9a + 3)$$
$$= 6a[2a(3a - 1) - 3(3a - 1)]$$
$$= 6a(3a - 1)(2a - 3)$$

103. $9p^2 + 30pq + 21q^2 = 3(3p^2 + 10pq + 7q^2)$
$$= 3(3p^2 + 7pq + 3pq + 7q^2)$$
$$= 3[p(3p + 7q) + q(3p + 7q)]$$
$$= 3(p + q)(3p + 7q)$$

105. $x^2 + kx + 8 \Rightarrow mn = 8$ and $m + n = k$
$$2 \cdot 4 = 8 \text{ and } m + n = 6$$
$$1 \cdot 8 = 8 \text{ and } m + n = 9$$
$k = 6$ or $k = 9$

107. $x^2 - kx + 16 \Rightarrow mn = 16$ and $m + n = -k$
$$-4 \cdot (-4) = 16 \text{ and } m + n = -8$$
$$-2 \cdot (-8) = 16 \text{ and } m + n = -10$$
$$-1 \cdot (-16) = 16 \text{ and } m + n = -17$$
$-k = -8, \ -k = -10, \ -k = -17$
$k = 8$ or $k = 10$ or $k = 17$

109. $x^2 - kx - 5 \Rightarrow mn = -5$ and $m + n = -k$
$$-1 \cdot 5 = -5 \text{ and } m + n = 4$$
$$1 \cdot (-5) = -5 \text{ and } m + n = -4$$
$-k = -4$ gives $k = 4$

111. $x^2 + 3x + k \Rightarrow m + n = 3$ and $mn = k$
$$1 + 2 = 3 \text{ and } mn = 2$$
$$2 + 1 = 3 \text{ and } mn = 2$$
$k = 2$

113. $x^2 + 2x - k = (x + m)(x - n)$
$$= x^2 + (m - n)x - mn$$

$m - n = 2$	$k = mn$
$3 - 1 = 2$	$k = 3 \cdot 1 = 3$
$4 - 2 = 2$	$k = 4 \cdot 2 = 8$
$5 - 3 = 2$	$k = 5 \cdot 3 = 15$
$6 - 4 = 2$	$k = 6 \cdot 4 = 24$

$k = 3, \ 8, \ 15, \ 24, \ \ldots$

Exercises 4.7

1. $(x - 3)(x - 4) = 0$
$x - 3 = 0$ or $x - 4 = 0$
$x = 3$ or $\quad x = 4$

3. $(3x + 1)(x - 6) = 0$
$3x + 1 = 0$ or $x - 6 = 0$
$x = -\dfrac{1}{3}$ or $\quad x = 6$

5. $x^2 - 2x - 3 = 0$
$(x - 3)(x + 1) = 0$
$x - 3 = 0$ or $x + 1 = 0$
$x = 3$ or $\quad x = -1$

7. $x^2 - 7x + 6 = 0$
$(x - 6)(x - 1) = 0$
$x - 6 = 0$ or $x - 1 = 0$
$x = 6$ or $\quad x = 1$

9. $x^2 + 8x + 15 = 0$
$(x + 5)(x + 3) = 0$
$x + 5 = 0$ or $x + 3 = 0$
$x = -5$ or $\quad x = -3$

11. $x^2 + 4x - 21 = 0$
$(x + 7)(x - 3) = 0$
$x + 7 = 0$ or $x - 3 = 0$
$x = -7$ or $\quad x = 3$

13. $x^2 - 4x = 12$
$x^2 - 4x - 12 = 0$
$(x - 6)(x + 2) = 0$
$x - 6 = 0$ or $x + 2 = 0$
$x = 6$ or $\quad x = -2$

15. $x^2 + 5x = 14$
$x^2 + 5x - 14 = 0$
$(x + 7)(x - 2) = 0$
$x + 7 = 0$ or $x - 2 = 0$
$x = -7$ or $\quad x = 2$

17. $2x^2 + 5x - 3 = 0$
$(2x - 1)(x + 3) = 0$
$2x - 1 = 0$ or $x + 3 = 0$
$x = \dfrac{1}{2}$ or $\quad x = -3$

19. $4x^2 - 24x + 35 = 0$
$(2x - 5)(2x - 7) = 0$
$2x - 5 = 0$ or $2x - 7 = 0$
$x = \dfrac{5}{2}$ or $\quad x = \dfrac{7}{2}$

21. $4x^2 + 11x = -6$
$4x^2 + 11x + 6 = 0$
$(4x + 3)(x + 2) = 0$
$4x + 3 = 0$ or $x + 2 = 0$
$x = -\dfrac{3}{4}$ or $\quad x = -2$

23. $5x^2 + 13x = 6$
$5x^2 + 13x - 6 = 0$
$(5x - 2)(x + 3) = 0$
$5x - 2 = 0$ or $x + 3 = 0$
$x = \dfrac{2}{5}$ or $\quad x = -3$

25. $x^2 - 2x = 0$
$x(x - 2) = 0$
$x = 0$ or $x - 2 = 0$
$\quad x = 2$

27. $x^2 = -8x$
$x^2 + 8x = 0$
$x(x + 8) = 0$
$x = 0$ or $x + 8 = 0$
$\quad x = -8$

29. $5x^2 - 15x = 0$

$5x(x - 3) = 0$

$5x = 0$ or $x - 3 = 0$

$x = 0$ or $x = 3$

31. $x^2 - 25 = 0$

$(x + 5)(x - 5) = 0$

$x + 5 = 0$ or $x - 5 = 0$

$x = -5$ or $x = 5$

33. $x^2 = 81$

$x^2 - 81 = 0$

$(x + 9)(x - 9) = 0$

$x + 9 = 0$ or $x - 9 = 0$

$x = -9$ or $x = 9$

35. $2x^2 - 18 = 0$

$2(x^2 - 9) = 0$

$2(x + 3)(x - 3) = 0$

$x + 3 = 0$ or $x - 3 = 0$

$x = -3$ or $x = 3$

37. $3x^2 + 24x + 45 = 0$

$3(x^2 + 8x + 15) = 0$

$3(x + 5)(x + 3) = 0$

$x + 5 = 0$ or $x + 3 = 0$

$x = -5$ or $x = -3$

39. $6x^2 + 28x = 10$

$6x^2 + 28x - 10 = 0$

$2(3x^2 + 14x - 5) = 0$

$2(3x - 1)(x + 5) = 0$

$3x - 1 = 0$ or $x + 5 = 0$

$x = \dfrac{1}{3}$ or $x = -5$

41. $(x + 3)(x - 2) = 14$

$x^2 + x - 6 = 14$

$x^2 + x - 20 = 0$

$(x + 5)(x - 4) = 0$

$x + 5 = 0$ or $x - 4 = 0$

$x = -5$ or $x = 4$

43. $x = 1^{st}$ integer

$x + 1 =$ next integer

$x(x + 1) = 132$

$x^2 + x - 132 = 0$

$(x + 12)(x - 11) = 0$

$x = -12$ or $x = 11$

$x + 1 = -11$ $x + 1 = 12$

The integers are -12, -11 or 11, 12.

45. $x =$ the integer

$x^2 + x = 72$

$x^2 + x - 72 = 0$

$(x + 9)(x - 8) = 0$

$x = -9$ or $x = 8$

The integer is -9 or 8.

47. $x =$ side length of original square.

new area $=$ original area $+ 39$

$(x + 3)^2 = x^2 + 39$

$x^2 + 6x + 9 = x^2 + 39$

$6x = 30$

$x = 5$

The original square was 5 in. by 5 in.

49. $P = -20$

$x^2 - 3x - 60 = -20$

$x^2 - 3x - 40 = 0$

$(x - 8)(x + 5) = 0$

$x = 8$ or $x = -5$

There were 8 appliances sold.

51. Writing exercise

Summary Exercises for Chapter 4

1. $18a + 24 = 6 \cdot 3a + 6 \cdot 4$

$= 6(3a + 4)$

3. $24s^2t - 16s^2 = 8s^2 \cdot 3t - 8s^2 \cdot 2$

$= 8s^2(3t - 2)$

5. $35s^3 - 28s^2 = 7s^2 \cdot 5s - 7s^2 \cdot 4$

$= 7s^2(5s - 4)$

7. $18m^2n^2 - 27m^2n + 45m^2n^3$

$= 9m^2n \cdot 2n - 9m^2n \cdot 3 + 9m^2n \cdot 5n^2$

$= 9m^2n(2n - 3 + 5n^2)$

$= 9m^2n(5n^2 + 2n - 3)$

$= 9m^2n(5n - 3)(n + 1)$

9. $8a^2b + 24ab - 16ab^2 = 8ab \cdot a + 8ab \cdot 3 - 8ab \cdot 2b$

$= 8ab(a + 3 - 2b)$

11. $x(2x - y) + y(2x - y) = (2x - y) \cdot x + (2x - y) \cdot y$

$= (2x - y)(x + y)$

13. $x^2 + 9x + 20$

Consider only the positive factors of 20 because the middle and last terms are positive.

$20 = 20 \cdot 1$

$= 10 \cdot 2$

$= 5 \cdot 4$

Possible Factors	Middle Terms
$(x + 20)(x + 1)$	$21x$
$(x + 10)(x + 2)$	$12x$
$(x + 5)(x + 4)$	$9x$

$x^2 + 9x + 20 = (x + 5)(x + 4)$

15. $a^2 - a - 12$

The factors of -12 will have different signs.

$-12 = (-12)(1)$

$= (12)(-1)$

$= (-6)(2)$

$= (6)(-2)$

$= (-4)(3)$

$= (4)(-3)$

Possible Factors	Middle Terms
$(a - 12)(a + 1)$	$-11a$
$(a + 12)(a - 1)$	$11a$
$(a - 6)(a + 2)$	$-4a$
$(a + 6)(a - 2)$	$4a$
$(a - 4)(a + 3)$	$-a$

$a^2 - a - 12 = (a - 4)(a + 3)$

17. $x^2 + 12x + 36$

Consider only the positive factors of 36 because the middle and last terms are positive.

$36 = 36 \cdot 1$

$= 18 \cdot 2$

$= 12 \cdot 3$

$= 9 \cdot 4$

$= 6 \cdot 6$

Possible Factors	Middle Terms
$(x + 36)(x + 1)$	$37x$
$(x + 18)(x + 2)$	$20x$
$(x + 12)(x + 3)$	$15x$
$(x + 9)(x + 4)$	$13x$
$(x + 6)(x + 6)$	$12x$

$x^2 + 12x + 36 = (x + 6)(x + 6)$

19. $b^2 - 4bc - 21c^2$

The factors of -21 will have different signs.

$-21 = (-21)(1)$

$= (21)(-1)$

$= (-7)(3)$

$= (7)(-3)$

Possible Factors	Middle Terms
$(b - 21c)(b + c)$	$-20bc$
$(b + 21c)(b - c)$	$20bc$
$(b - 7c)(b + 3c)$	$-4bc$

$b^2 - 4bc - 21c^2 = (b - 7c)(b + 3c)$

21. $m^3 + 2m^2 - 35m = m(m^2 + 2m - 35)$

The factors of -35 will have different signs.

$-35 = (-35)(1)$

$= (35)(-1)$

$= (-7)(5)$

$= (7)(-5)$

Possible Factors	Middle Terms
$(m - 35)(m + 1)$	$-34m$
$(m + 35)(m - 1)$	$34m$
$(m - 7)(m + 5)$	$-2m$
$(m + 7)(m - 5)$	$2m$

$m^3 + 2m^2 - 35m = m(m + 7)(m - 5)$

23. $3y^3 - 48y^2 + 189y = 3y(y^2 - 16y + 63)$

Consider only the negative factors of 63 because the middle term is negative.

$$63 = (-63)(-1)$$
$$= (-21)(-3)$$
$$= (-9)(-7)$$

Possible Factors	Middle Terms
$(y - 63)(y - 1)$	$-64y$
$(y - 21)(y - 3)$	$-24y$
$(y - 9)(y - 7)$	$-16y$

$3y^3 - 48y^2 + 189y = 3y(y - 9)(y - 7)$

25. $3x^2 + 8x + 5$

$$3 = 3 \cdot 1 \qquad 5 = 5 \cdot 1$$

Consider only the positive factors of 5 because the middle term is positive.

Possible Factors	Middle Terms
$(3x + 1)(x + 5)$	$16x$
$(3x + 5)(x + 1)$	$8x$

$3x^2 + 8x + 5 = (3x + 5)(x + 1)$

27. $2b^2 - 9b + 9$

$$2 = 2 \cdot 1 \qquad 9 = 9 \cdot 1$$
$$= 3 \cdot 3$$

Consider only the negative factors of 9 because the middle term is negative.

Possible Factors	Middle Terms
$(2b - 9)(b - 1)$	$-11b$
$(2b - 1)(b - 9)$	$-19b$
$(2b - 3)(b - 3)$	$-9b$

$2b^2 - 9b + 9 = (2b - 3)(b - 3)$

29. $10x^2 - 11x + 3$

$$10 = 10 \cdot 1 \qquad 3 = 3 \cdot 1$$
$$= 5 \cdot 2$$

Consider only the negative factors of 3 because the middle term is negative.

Possible Factors	Middle Terms
$(10x - 3)(x - 1)$	$-13x$
$(10x - 1)(x - 3)$	$-31x$
$(5x - 3)(2x - 1)$	$-11x$

$10x^2 - 11x + 3 = (5x - 3)(2x - 1)$

31. $9y^2 - 3yz - 20z^2$

$$9 = 3 \cdot 3 \qquad 20 = 5 \cdot 4$$
$$= 9 \cdot 1 \qquad = 10 \cdot 2$$
$$= 20 \cdot 1$$

Only one factor will have a negative sign because the last term is negative.

Possible Factors	Middle Terms
$(3y + 5z)(3y - 4z)$	$3yz$
$(3y - 5z)(3y + 4z)$	$-3yz$

$9y^2 - 3yz - 20z^2 = (3y - 5z)(3y + 4z)$

33. $8x^3 - 36x^2 - 20x = 4x(2x^2 - 9x - 5)$

$$2 = 1 \cdot 2 \qquad 5 = 1 \cdot 5$$

Only one factor will have a negative sign because the last term is negative.

Possible Factors	Middle Terms
$(x + 1)(2x - 5)$	$-3x$
$(x - 1)(2x + 5)$	$3x$
$(2x + 1)(x - 5)$	$-9x$

$8x^3 - 36x^2 - 20x = 4x(2x + 1)(x - 5)$

35. $6x^3 - 3x^2 - 9x = 3x(2x^2 - x - 3)$

$$2 = 1 \cdot 2 \qquad 3 = 1 \cdot 3$$

Only one factor will have a negative sign because the last term is negative.

Possible Factors	Middle Terms
$(x - 1)(2x + 3)$	x
$(x + 1)(2x - 3)$	$-x$

$6x^3 - 3x^2 - 9x = 3x(x + 1)(2x - 3)$

37. $p^2 - 49 = (p + 7)(p - 7)$

39. $m^2 - 9n^2 = (m + 3n)(m - 3n)$

41. $25 - z^2 = (5 + z)(5 - z)$

43. $25a^2 - 36b^2 = (5a + 6b)(5a - 6b)$

45. $3w^3 - 12wz^2 = 3w(w^2 - 4z^2)$
$\qquad\qquad\quad = 3w(w + 2z)(w - 2z)$

47. $2m^2 - 72n^4 = 2(m^2 - 36n^4)$
$\qquad\qquad\quad = 2(m + 6n^2)(m - 6n^2)$

49. $x^2 + 8x + 16 = (x + 4)(x + 4)$
$\qquad\qquad\qquad = (x + 4)^2$

51. $4x^2 + 12x + 9 = (2x + 3)(2x + 3)$
$\qquad\qquad\qquad\; = (2x + 3)^2$

53. $16x^3 + 40x^2 + 25x = x(16x^2 + 40x + 25)$
$\qquad\qquad\qquad\qquad\; = x(4x + 5)^2$

55. $x^2 - 4x + 5x - 20 = (x^2 - 4x) + (5x - 20)$
$\qquad\qquad\qquad\qquad = x(x - 4) + 5(x - 4)$
$\qquad\qquad\qquad\qquad = (x - 4)(x + 5)$

57. $6x^2 + 4x - 15x - 10 = (6x^2 + 4x) + (-15x - 10)$
$\qquad\qquad\qquad\qquad\;\; = 2x(3x + 2) - 5(3x + 2)$
$\qquad\qquad\qquad\qquad\;\; = (3x + 2)(2x - 5)$

59. $6x^3 + 9x^2 - 4x^2 - 6x = x(6x^2 + 9x - 4x - 6)$
$\qquad\qquad\qquad\qquad\quad = x[(6x^2 + 9x) + (-4x - 6)]$
$\qquad\qquad\qquad\qquad\quad = x[3x(2x + 3) - 2(2x + 3)]$
$\qquad\qquad\qquad\qquad\quad = x(2x + 3)(3x - 2)$

61. $(x - 1)(2x + 3) = 0$
$\quad\; x - 1 = 0 \;$ or $\; 2x + 3 = 0$
$\qquad\; x = 1 \;$ or $\qquad x = -\dfrac{3}{2}$

63. $x^2 - 10x = 0$
$\quad\; x(x - 10) = 0$
$\quad\; x = 0 \;$ or $\; x - 10 = 0$
$\qquad\qquad\qquad\; x = 10$

65. $\qquad\; x^2 - 2x = 15$
$\qquad x^2 - 2x - 15 = 0$
$\qquad (x + 3)(x - 5) = 0$
$\quad\; x + 3 = 0 \quad$ or $\; x - 5 = 0$
$\qquad\; x = -3 \;$ or $\qquad x = 5$

67. $4x^2 - 13x + 10 = 0$
$\quad\; (x - 2)(4x - 5) = 0$
$\quad\; x - 2 = 0 \;$ or $\; 4x - 5 = 0$
$\qquad\; x = 2 \;$ or $\qquad x = \dfrac{5}{4}$

69. $3x^2 - 9x = 0$
$\quad\; 3x(x - 3) = 0$
$\quad\; 3x = 0 \;$ or $\; x - 3 = 0$
$\qquad\; x = 0 \;$ or $\qquad x = 3$

71. $\qquad\quad 2x^2 - 32 = 0$
$\qquad\quad\; 2(x^2 - 16) = 0$
$\qquad 2(x + 4)(x - 4) = 0$
$\quad\; x + 4 = 0 \quad$ or $\; x - 4 = 0$
$\qquad\; x = -4 \;$ or $\qquad x = 4$

Exercises 5.1

1. $\dfrac{16}{24} = \dfrac{2 \cdot 2 \cdot 2 \cdot 2}{2 \cdot 2 \cdot 2 \cdot 3} = \dfrac{2}{3}$

3. $\dfrac{80}{180} = \dfrac{2 \cdot 2 \cdot 2 \cdot 2 \cdot 5}{2 \cdot 2 \cdot 3 \cdot 3 \cdot 5} = \dfrac{4}{9}$

5. $\dfrac{4x^5}{6x^2} = \dfrac{2 \cdot 2 \cdot x \cdot x \cdot x \cdot x \cdot x}{2 \cdot 3 \cdot x \cdot x} = \dfrac{2x^3}{3}$

7. $\dfrac{9x^3}{27x^6} = \dfrac{3 \cdot 3 \cdot x \cdot x \cdot x}{3 \cdot 3 \cdot 3 \cdot x \cdot x \cdot x \cdot x \cdot x \cdot x} = \dfrac{1}{3x^3}$

9. $\dfrac{10a^2 b^5}{25ab^2} = \dfrac{2 \cdot 5 \cdot a \cdot a \cdot b \cdot b \cdot b \cdot b \cdot b}{5 \cdot 5 \cdot a \cdot b \cdot b} = \dfrac{2ab^3}{5}$

11. $\dfrac{42x^3 y}{14xy^3} = \dfrac{2 \cdot 3 \cdot 7 \cdot x \cdot x \cdot x \cdot y}{2 \cdot 7 \cdot x \cdot y \cdot y \cdot y} = \dfrac{3x^2}{y^2}$

13. $\dfrac{2xyw^2}{6x^2 y^3 w^3} = \dfrac{2 \cdot x \cdot y \cdot w \cdot w}{2 \cdot 3 \cdot x \cdot x \cdot y \cdot y \cdot y \cdot w \cdot w \cdot w} = \dfrac{1}{3xy^2 w}$

15. $\dfrac{10x^5 y^5}{2x^3 y^4} = \dfrac{2 \cdot 5 \cdot x \cdot x \cdot x \cdot x \cdot x \cdot y \cdot y \cdot y \cdot y \cdot y}{2 \cdot x \cdot x \cdot x \cdot y \cdot y \cdot y \cdot y}$

$\qquad = 5x^2 y$

17. $\dfrac{-4m^3 n}{6mn^2} = \dfrac{-2 \cdot 2 \cdot m \cdot m \cdot m \cdot n}{2 \cdot 3 \cdot m \cdot n \cdot n} = \dfrac{-2m^2}{3n}$

19. $\dfrac{-8ab^3}{-16a^3 b} = \dfrac{-2 \cdot 2 \cdot 2 \cdot a \cdot b \cdot b \cdot b}{-2 \cdot 2 \cdot 2 \cdot 2 \cdot a \cdot a \cdot a \cdot b} = \dfrac{b^2}{2a^2}$

21. $\dfrac{8r^2 s^3 t}{-16rs^4 t^3} = \dfrac{2 \cdot 2 \cdot 2 \cdot r \cdot r \cdot s \cdot s \cdot s \cdot t}{-2 \cdot 2 \cdot 2 \cdot 2 \cdot r \cdot s \cdot s \cdot s \cdot s \cdot t \cdot t \cdot t}$

$\qquad = \dfrac{-r}{2st^2}$

23. $\dfrac{3x + 18}{5x + 30} = \dfrac{3(x+6)}{5(x+6)} = \dfrac{3}{5}$

25. $\dfrac{3x - 6}{5x - 15} = \dfrac{3(x-2)}{5(x-3)}$

27. $\dfrac{6a - 24}{a^2 - 16} = \dfrac{6(a-4)}{(a+4)(a-4)} = \dfrac{6}{a+4}$

29. $\dfrac{x^2 + 3x + 2}{5x + 10} = \dfrac{(x+2)(x+1)}{5(x+2)} = \dfrac{x+1}{5}$

31. $\dfrac{x^2 - 6x - 16}{x^2 - 64} = \dfrac{(x-8)(x+2)}{(x+8)(x-8)} = \dfrac{x+2}{x+8}$

33. $\dfrac{2m^2 + 3m - 5}{2m^2 + 11m + 15} = \dfrac{(2m+5)(m-1)}{(2m+5)(m+3)} = \dfrac{m-1}{m+3}$

35. $\dfrac{p^2 + 2pq - 15q^2}{p^2 - 25q^2} = \dfrac{(p+5q)(p-3q)}{(p+5q)(p-5q)} = \dfrac{p-3q}{p-5q}$

37. $\dfrac{2x - 10}{25 - x^2} = \dfrac{2(x-5)}{(5+x)(5-x)}$

$\qquad = \dfrac{2(x-5)}{-(x+5)(x-5)}$

$\qquad = \dfrac{-2}{x+5}$

39. $\dfrac{25 - a^2}{a^2 + a - 30} = \dfrac{(5+a)(5-a)}{(a+6)(a-5)} = \dfrac{-(a+5)}{a+6} = \dfrac{-a-5}{x+5}$

41. $\dfrac{x^2 + xy - 6y^2}{4y^2 - x^2} = \dfrac{(x+3y)(x-2y)}{(2y+x)(2y-x)}$

$\qquad = \dfrac{-(x+3y)}{x+2y}$

$\qquad = \dfrac{-x-3y}{x+2y}$

43. $\dfrac{x^2 + 4x + 4}{x+2} = \dfrac{(x+2)(x+2)}{x+2} = x+2$

45. $\dfrac{xy - 2y + 4x - 8}{2y + 6 - xy - 3x} = \dfrac{y(x-2) + 4(x-2)}{2(y+3) - x(y+3)}$

$\qquad = \dfrac{(x-2)(y+4)}{(y+3)(2-x)}$

$\qquad = \dfrac{-(y+4)}{y+3}$

47. $\dfrac{y-7}{7-y} = \dfrac{y-7}{-(y-7)} = -1$

49. $\dfrac{6x^2 + 19x + 10}{3x + 2} = \dfrac{(3x+2)(2x+5)}{3x+2}$

$\qquad = 2x + 5$

The length is represented by $2x + 5$.

51. Writing exercise

53. Writing exercise

55. Writing exercise

Exercises 5.2

1. $\dfrac{7}{18} + \dfrac{5}{18} = \dfrac{7+5}{18} = \dfrac{12}{18} = \dfrac{2}{3}$

3. $\dfrac{13}{16} - \dfrac{9}{16} = \dfrac{13-9}{16} = \dfrac{4}{16} = \dfrac{1}{4}$

5. $\dfrac{x}{8} + \dfrac{3x}{8} = \dfrac{x+3x}{8} = \dfrac{4x}{8} = \dfrac{x}{2}$

7. $\dfrac{7a}{10} - \dfrac{3a}{10} = \dfrac{7a-3a}{10} = \dfrac{4a}{10} = \dfrac{2a}{5}$

9. $\dfrac{5}{x} + \dfrac{3}{x} = \dfrac{5+3}{x} = \dfrac{8}{x}$

11. $\dfrac{8}{w} - \dfrac{2}{w} = \dfrac{8-2}{w} = \dfrac{6}{w}$

13. $\dfrac{2}{xy} + \dfrac{3}{xy} = \dfrac{2+3}{xy} = \dfrac{5}{xy}$

15. $\dfrac{2}{3cd} + \dfrac{4}{3cd} = \dfrac{2+4}{3cd} = \dfrac{6}{3cd} = \dfrac{2}{cd}$

17. $\dfrac{7}{x-5} + \dfrac{9}{x-5} = \dfrac{7+9}{x-5} = \dfrac{16}{x-5}$

19. $\dfrac{2x}{x-2} - \dfrac{4}{x-2} = \dfrac{2x-4}{x-2} = \dfrac{2(x-2)}{x-2} = 2$

21. $\dfrac{8p}{p+4} + \dfrac{32}{p+4} = \dfrac{8p+32}{p+4} = \dfrac{8(p+4)}{p+4} = 8$

23. $\dfrac{x^2}{x-2} - \dfrac{4}{x-2} = \dfrac{2x-4}{x-2} = \dfrac{2(x-2)}{x-2} = 2$

25. $\dfrac{m^2}{m-5} - \dfrac{25}{m-5} = \dfrac{m^2-25}{m-5}$
$= \dfrac{(m+5)(m-5)}{m-5}$
$= m+5$

27. $\dfrac{a-1}{3} + \dfrac{2a-5}{3} = \dfrac{(a-1)+(2a-5)}{3}$
$= \dfrac{a-1+2a-5}{3}$
$= \dfrac{3a-6}{3} = \dfrac{3(a-2)}{3}$
$= a-2$

29. $\dfrac{3x-1}{4} - \dfrac{x+7}{4} = \dfrac{(3x-1)-(x+7)}{4}$
$= \dfrac{y+2+4y+8}{5}$
$= \dfrac{5y+10}{5}$
$= \dfrac{5(y+2)}{5}$
$= y+2$

31. $\dfrac{4m+7}{6m} + \dfrac{2m+5}{6m} = \dfrac{(4m+7)+(2m+5)}{6m}$
$= \dfrac{4m+7+2m+5}{6m}$
$= \dfrac{6m+12}{6m}$
$= \dfrac{6(m+2)}{6m} =$
$\dfrac{m+2}{m}$

33. $\dfrac{4w-y}{w-5} - \dfrac{2w+3}{w-5} = \dfrac{(4w-7)-(2w+3)}{w-5}$
$= \dfrac{4w-7-2w-3}{w-5}$
$= \dfrac{2w-10}{w-5}$
$= \dfrac{2(w-5)}{w-5}$
$= 2$

35. $\dfrac{x-7}{x^2-x-6} + \dfrac{2x-2}{x^2-x-6} = \dfrac{(x-7)+(2x-2)}{x^2-x-6}$
$= \dfrac{x-y+2x-2}{x^2-x-6}$
$= \dfrac{3x-9}{(x-3)(x+2)}$
$= \dfrac{3(x-3)}{(x-3)9x+2)}$
$= \dfrac{3}{x+2}$

37. $\dfrac{y^2}{2y+8} + \dfrac{3y-4}{2y+8} = \dfrac{y^2+3y-4}{2y+8}$

$\qquad\qquad\qquad = \dfrac{(y+4)(y-1)}{2(y+4)}$

$\qquad\qquad\qquad = \dfrac{y-1}{2}$

39. $\dfrac{7w}{w+3} + \dfrac{21}{w+3} = \dfrac{7w+21}{w+3} = \dfrac{7(w+3)}{w+3} = 7$

41. $\dfrac{x^2}{x^2+x-6} - \dfrac{6}{(x+3)(x-2)} + \dfrac{x}{x^2+x-6} = \dfrac{x^2}{(x+3)(x-2)} - \dfrac{6}{(x+3)(x-2)} + \dfrac{x}{(x+3)(x-2)}$

$\qquad\qquad\qquad\qquad\qquad\qquad = \dfrac{x^2-6+x}{(x+3)(x-2)}$

$\qquad\qquad\qquad\qquad\qquad\qquad = \dfrac{(x+3)(x-2)}{(x+3)(x-2)}$

$\qquad\qquad\qquad\qquad\qquad\qquad = 1$

43. $\dfrac{2x}{x+3} + \dfrac{2x}{x+3} + \dfrac{6}{x+3} + \dfrac{6}{x+3} = \dfrac{2x+2x+6+6}{x+3}$

$\qquad\qquad\qquad\qquad\qquad\qquad = \dfrac{4x+12}{x+3}$

$\qquad\qquad\qquad\qquad\qquad\qquad = \dfrac{4(x+3)}{x+3}$

$\qquad\qquad\qquad\qquad\qquad\qquad = 4$

Exercises 5.3

1. LCD: $7 \cdot 6 = 42$

$\dfrac{3}{7} + \dfrac{5}{6} = \dfrac{3 \cdot 6}{7 \cdot 6} + \dfrac{5 \cdot 7}{6 \cdot 7} = \dfrac{18}{42} + \dfrac{35}{42} = \dfrac{18+35}{42} = \dfrac{53}{42}$

3. $25 = 5 \cdot 5$

$20 = 5 \cdot 2 \cdot 2$

LCD: $5 \cdot 5 \cdot 2 \cdot 2 = 100$

$\dfrac{13}{25} - \dfrac{7}{20} = \dfrac{13 \cdot 4}{25 \cdot 4} - \dfrac{7 \cdot 5}{20 \cdot 5} = \dfrac{52}{100} - \dfrac{35}{100} = \dfrac{52-35}{100} = \dfrac{17}{100}$

5. LCD: $4 \cdot 5 = 20$

$\dfrac{y}{4} + \dfrac{3y}{5} = \dfrac{y \cdot 5}{4 \cdot 5} + \dfrac{3y \cdot 4}{5 \cdot 4} = \dfrac{5y}{20} + \dfrac{12y}{20} = \dfrac{5y+12y}{20} = \dfrac{17y}{20}$

7. LCD: $7 \cdot 3 = 21$

$\dfrac{7a}{3} - \dfrac{a}{7} = \dfrac{7a \cdot 7}{3 \cdot 7} - \dfrac{a \cdot 3}{7 \cdot 3} = \dfrac{49a}{21} - \dfrac{3a}{21} = \dfrac{49a-3a}{21} = \dfrac{46a}{21}$

9. LCD: $x \cdot 5 = 5x$

$$\frac{3}{x} - \frac{4}{5} = \frac{3 \cdot 5}{x \cdot 5} - \frac{4 \cdot x}{5 \cdot x} = \frac{15}{5x} - \frac{4x}{5x} = \frac{15 - 4x}{5x}$$

11. LCD: $a \cdot 5 = 5a$

$$\frac{5}{a} + \frac{a}{5} = \frac{5 \cdot 5}{a \cdot 5} + \frac{a \cdot a}{5 \cdot a} = \frac{25}{5a} + \frac{a^2}{5a} = \frac{25 + a^2}{5a}$$

13. $m^2 = m \cdot m$

LCD: m^2

$$\frac{5}{m} + \frac{3}{m^2} = \frac{5 \cdot m}{m \cdot m} + \frac{3}{m^2} = \frac{5m}{m^2} + \frac{3}{m^2} = \frac{5m + 3}{m^2}$$

15. $x^2 = x \cdot x$

$7x = 7 \cdot x$

LCD: $7 \cdot x \cdot x = 7x^2$

$$\frac{2}{x^2} - \frac{5}{7x} = \frac{2 \cdot 7}{x^2 \cdot 7} - \frac{5 \cdot x}{7x \cdot x}$$
$$= \frac{14}{7x^2} - \frac{5x}{7x^2}$$
$$= \frac{14 - 5x}{7x^2}$$

17. $9s = 9 \cdot s$

$s^2 = s \cdot s$

LCD: $9 \cdot s \cdot s = 9s^2$

$$\frac{7}{9s} + \frac{5}{s^2} = \frac{7 \cdot s}{9s \cdot s} + \frac{5 \cdot 9}{s^2 \cdot 9} = \frac{7s}{9s^2} + \frac{45}{9s^2} = \frac{7s + 45}{9s^2}$$

19. $4b^2 = 2 \cdot 2 \cdot b \cdot b$

$3b^3 = 3 \cdot b \cdot b \cdot b$

LCD: $2 \cdot 2 \cdot 3 \cdot b \cdot b \cdot b = 12b^3$

$$\frac{3}{4b^2} + \frac{5}{3b^3} = \frac{3 \cdot 3b}{4b^2 \cdot 3b} + \frac{5 \cdot 4}{3b^3 \cdot 4}$$
$$= \frac{9b}{12b^3} + \frac{20}{12b^3}$$
$$= \frac{9b + 20}{12b^3}$$

21. LCD: $5(x + 2)$

$$\frac{x}{x+2} + \frac{2}{5} = \frac{x \cdot 5}{(x+2) \cdot 5} + \frac{2(x+2)}{5(x+2)}$$
$$= \frac{5x}{5(x+2)} + \frac{2(x+2)}{5(x+2)}$$
$$= \frac{5x + 2(x+2)}{5(x+2)}$$
$$= \frac{5x + 2x + 4}{5(x+2)}$$
$$= \frac{7x + 4}{5(x+2)}$$

23. LCD: $4(y - 4)$

$$\frac{y}{y-4} - \frac{3}{4} = \frac{y \cdot 4}{(y-4) \cdot 4} - \frac{3(y-4)}{4(y-4)}$$
$$= \frac{4y - 3(y-4)}{4(y-4)}$$
$$= \frac{4y - 3y + 12}{4(y-4)}$$
$$= \frac{y + 12}{4(y-4)}$$

25. LCD: $x(x + 1)$

$$\frac{4}{x} + \frac{3}{x+1} = \frac{4 \cdot (x+1)}{x \cdot (x+1)} + \frac{3 \cdot x}{(x+1) \cdot x}$$
$$= \frac{4(x+1)}{x(x+1)} + \frac{3x}{x(x+1)}$$
$$= \frac{4(x+1) + 3x}{x(x+1)}$$
$$= \frac{4x + 4 + 3x}{x(x+1)}$$
$$= \frac{7x + 4}{x(x+1)}$$

27. LCD: $a(a - 1)$

$$\frac{5}{a-1} - \frac{2}{a} = \frac{5 \cdot a}{(a-1) \cdot a} - \frac{2 \cdot (a-1)}{a \cdot (a-1)}$$
$$= \frac{5a}{a(a-1)} - \frac{2(a-1)}{a(a-1)}$$
$$= \frac{5a - 2(a-1)}{a(a-1)}$$
$$= \frac{5a - 2a + 2}{a(a-1)}$$
$$= \frac{3a + 2}{a(a-1)}$$

29. LCD: $(2x - 3)(3x)$

$$\frac{4}{2x-3} + \frac{2}{3x} = \frac{4 \cdot 3x}{(2x-3) \cdot 3x} + \frac{2 \cdot (2x-3)}{3x \cdot (2x-3)}$$
$$= \frac{12x}{3x(2x-3)} + \frac{2(2x-3)}{3x(2x-3)}$$
$$= \frac{12x + 2(2x-3)}{3x(2x-3)}$$
$$= \frac{12x + 4x - 6}{3x(2x-3)}$$
$$= \frac{16x - 6}{3x(2x-3)}$$
$$= \frac{2(8x - 3)}{3x(2x-3)}$$

31. LCD: $(x+1)(x+3)$

$$\frac{2}{x+1}+\frac{3}{x+3}=\frac{2\cdot(x+3)}{(x+1)\cdot(x+3)}+\frac{3\cdot(x+1)}{(x+3)\cdot(x+1)}$$

$$=\frac{2(x+3)}{(x+1)(x+3)}+\frac{3(x+1)}{(x+1)(x+3)}$$

$$=\frac{2(x+3)+3(x+1)}{(x+1)(x+3)}$$

$$=\frac{2x+6+3x+3}{(x+1)(x+3)}$$

$$=\frac{5x+9}{(x+1)(x+3)}$$

33. LCD: $(y-2)(y+1)$

$$\frac{4}{y-2}-\frac{1}{y+1}=\frac{4\cdot(y+1)}{(y-2)\cdot(y+1)}-\frac{1\cdot(y-2)}{(y+1)\cdot(y-2)}$$

$$=\frac{4(y+1)}{(y-2)(y+1)}-\frac{(y-2)}{(y-2)(y+1)}$$

$$=\frac{4(y+1)-(y-2)}{(y-2)(y+1)}$$

$$=\frac{4y+4-y+2}{(y-2)(y+1)}$$

$$=\frac{3y+6}{(y-2)(y+1)}$$

$$=\frac{3(y+2)}{(y-2)(y+1)}$$

35. $2b-6=2(b-3)$

LCD: $2(b-3)$

$$\frac{2}{b-3}+\frac{3}{2b-6}=\frac{2\cdot2}{(b-3)\cdot2}+\frac{3}{2(b-3)}$$

$$=\frac{4}{2(b-3)}+\frac{3}{2(b-3)}$$

$$=\frac{4+3}{2(b-3)}$$

$$=\frac{7}{2(b-3)}$$

37. $3x+12=3(x+4)$

LCD: $3(x+4)$

$$\frac{x}{x+4}-\frac{2}{3x+12}=\frac{x\cdot3}{(x+4)\cdot3}-\frac{2}{3(x+4)}$$

$$=\frac{3x}{3(x+4)}-\frac{2}{3(x+4)}$$

$$=\frac{3x-2}{3(x+4)}$$

39. $3m+3=3(m+1)$

$2m+2=2(m+1)$

LCD: $3\cdot2(m+1)=6(m+1)$

$$\frac{4}{3m+3}+\frac{1}{2m+2}=\frac{4}{3(m+1)}+\frac{1}{2(m+1)}$$

$$=\frac{4\cdot2}{3(m+1)\cdot2}+\frac{1\cdot3}{2(m+1)\cdot3}$$

$$=\frac{8}{6(m+1)}+\frac{3}{6(m+1)}$$

$$=\frac{8+3}{6(m+1)}$$

$$=\frac{11}{6(m+1)}$$

41. $5x-10=5(x-2)$

$3x-6=3(x-2)$

LCD: $5\cdot3(x-2)=15(x-2)$

$$\frac{4}{5x-10}-\frac{1}{3x-6}=\frac{4}{5(x-2)}-\frac{1}{3(x-2)}$$

$$=\frac{4\cdot3}{5(x-2)\cdot3}-\frac{1\cdot5}{3(x-2)\cdot5}$$

$$=\frac{12}{15(x-2)}-\frac{5}{15(x-2)}$$

$$=\frac{12-5}{15(x-2)}$$

$$=\frac{7}{15(x-2)}$$

43. $3c+6=3(c+2)$

$7c+14=7(c+2)$

LCD: $3\cdot7(c+2)=21(c+2)$

$$\frac{7}{3c+6}-\frac{2c}{7c+14}=\frac{7}{3(c+2)}-\frac{2c}{7(c+2)}$$

$$=\frac{7\cdot7}{3(c+2)\cdot7}-\frac{2c\cdot3}{7(c+2)\cdot3}$$

$$=\frac{49}{21(c+2)}-\frac{6c}{21(c+2)}$$

$$=\frac{49-6c}{21(c+2)}$$

45. $3y + 3 = 3(y + 1)$
LCD: $3(y + 1)$

$$\frac{y-1}{y+1} - \frac{y}{3y+3} = \frac{(y-1)\cdot 3}{(y+1)\cdot 3} - \frac{y}{3(y+1)}$$
$$= \frac{3(y-1)}{3(y+1)} - \frac{y}{3(y+1)}$$
$$= \frac{3(y-1) - y}{3(y+1)}$$
$$= \frac{3y - 3 - y}{3(y+1)}$$
$$= \frac{2y - 3}{3(y+1)}$$

47. $x^2 - 4 = (x+2)(x-2)$
LCD: $(x+2)(x-2)$

$$\frac{3}{x^2-4} + \frac{2}{x+2} = \frac{3}{(x+2)(x-2)} + \frac{2\cdot(x-2)}{(x+2)\cdot(x-2)}$$
$$= \frac{3 + 2(x-2)}{(x+2)(x-2)}$$
$$= \frac{3 + 2x - 4}{(x+2)(x-2)}$$
$$= \frac{2x - 1}{(x+2)(x-2)}$$

49. $x^2 - 3x + 2 = (x-1)(x-2)$
LCD: $(x-1)(x-2)$

$$\frac{3x}{x^2-3x+2} - \frac{1}{x-2}$$
$$= \frac{3x}{(x-1)(x-2)} - \frac{1\cdot(x-1)}{(x-2)\cdot(x-1)}$$
$$= \frac{3x - (x-1)}{(x-1)(x-2)}$$
$$= \frac{3x - x + 1}{(x-1)(x-2)}$$
$$= \frac{2x + 1}{(x-1)(x-2)}$$

51. $x^2 - 5x + 6 = (x-2)(x-3)$
LCD: $(x-2)(x-3)$

$$\frac{2x}{x^2-5x+6} + \frac{4}{x-2}$$
$$= \frac{2x}{(x-2)(x-3)} + \frac{4\cdot(x-3)}{(x-2)\cdot(x-3)}$$
$$= \frac{2x + 4(x-3)}{(x-2)(x-3)}$$
$$= \frac{2x + 4x - 12}{(x-2)(x-3)}$$
$$= \frac{6x - 12}{(x-2)(x-3)}$$
$$= \frac{6(\overset{1}{\cancel{x-2}})}{(\cancel{x-2})(x-3)}$$
$$= \frac{6}{x-3}$$

53. $3x - 3 = 3(x-1)$
$4x + 4 = 4(x+1)$
LCD: $3 \cdot 4(x-1)(x+1) = 12(x-1)(x+1)$

$$\frac{2}{3x-3} - \frac{1}{4x+4}$$
$$= \frac{2}{3(x-1)} - \frac{1}{4(x+1)}$$
$$= \frac{2\cdot 4(x+1)}{3(x-1)\cdot 4(x+1)} - \frac{1\cdot 3(x-1)}{4(x+1)\cdot 3(x-1)}$$
$$= \frac{8(x+1)}{12(x-1)(x+1)} - \frac{3(x-1)}{12(x-1)(x+1)}$$
$$= \frac{8(x+1) - 3(x-1)}{12(x-1)(x+1)}$$
$$= \frac{8x + 8 - 3x + 3}{12(x-1)(x+1)}$$
$$= \frac{5x + 11}{12(x-1)(x+1)}$$

55. $3a - 9 = 3(a - 3)$

$2a + 4 = 2(a + 2)$

LCD: $3 \cdot 2(a-3)(a+2) = 6(a-3)(a+2)$

$$\frac{4}{3a-9} - \frac{3}{2a+4}$$

$$= \frac{4}{3(a-3)} - \frac{3}{2(a+2)}$$

$$= \frac{4 \cdot 2(a+2)}{3(a-3) \cdot 2(a+2)} - \frac{3 \cdot 3(a-3)}{2(a+2) \cdot 3(a-3)}$$

$$= \frac{8(a+2)}{6(a-3)(a+2)} - \frac{9(a-3)}{6(a-3)(a+2)}$$

$$= \frac{8a + 16 - 9a + 27}{6(a-3)(a+2)}$$

$$= \frac{-a + 43}{6(a-3)(a+2)}$$

57. $x^2 - 16 = (x+4)(x-4)$

$x^2 - x - 12 = (x-4)(x+3)$

LCD: $(x+4)(x-4)(x+3)$

$$\frac{5}{x^2 - 16} - \frac{3}{x^2 - x - 12}$$

$$= \frac{5}{(x+4)(x-4)} - \frac{3}{(x-4)(x+3)}$$

$$= \frac{5 \cdot (x+3)}{(x+4)(x-4) \cdot (x+3)} - \frac{3 \cdot (x+4)}{(x-4)(x+3) \cdot (x+4)}$$

$$= \frac{5(x+3) - 3(x+4)}{(x+4)(x-4)(x+3)}$$

$$= \frac{5x + 15 - 3x - 12}{(x+4)(x-4)(x+3)}$$

$$= \frac{2x + 3}{(x+4)(x-4)(x+3)}$$

59. $y^2 + y - 6 = (y+3)(y-2)$

$y^2 - 2y - 15 = (y+3)(y-5)$

LCD: $(y+3)(y-2)(y-5)$

$$\frac{2}{y^2 + y - 6} + \frac{3y}{y^2 - 2y - 15}$$

$$= \frac{2}{(y+3)(y-2)} + \frac{3y}{(y-5)(y+3)}$$

$$= \frac{2 \cdot (y-5)}{(y+3)(y-2) \cdot (y-5)} + \frac{3y \cdot (y-2)}{(y-4)(y+3) \cdot (y-2)}$$

$$= \frac{2(y-5) + 3y(y-2)}{(y+3)(y-2)(y-5)}$$

$$= \frac{2y - 10 + 3y^2 - 6y}{(y+3)(y-2)(y-5)}$$

$$= \frac{3y^2 - 4y - 10}{(y+3)(y-2)(y-5)}$$

61. $x^2 - 9 = (x+3)(x-3)$

$x^2 + x - 6 = (x+3)(a-2)$

LCD: $(x+3)(x-3)(x-2)$

$$\frac{6x}{x^2 - 9} - \frac{5x}{x^2 + x - 6}$$

$$= \frac{6x}{(x-3)(x+3)} - \frac{5x}{(x+3)(x-2)}$$

$$= \frac{6x \cdot (x-2)}{(x-3)(x+3) \cdot (x-2)} - \frac{5x \cdot (x-3)}{(x+3)(x-2) \cdot (x-3)}$$

$$= \frac{6x(x-2) - 5x(x-3)}{(x-3)(x+3)(x-2)}$$

$$= \frac{6x^2 - 12x - 5x^2 + 15x}{(x-3)(x+3)(x-2)}$$

$$= \frac{x^2 + 3x}{(x-3)(x+3)(x-2)}$$

$$= \frac{x \overset{1}{\cancel{(x+3)}}}{(x-3)\underset{1}{\cancel{(x+3)}}(x-2)}$$

$$= \frac{x}{(x-3)(x-2)}$$

63. $\dfrac{3}{a-7} + \dfrac{2}{7-a} = \dfrac{3}{a-7} + \dfrac{2}{-(a-7)}$

$$= \frac{3-2}{a-7}$$

$$= \frac{1}{a-7}$$

65. $\dfrac{2x}{2x-3} - \dfrac{1}{3-2x} = \dfrac{2x}{2x-3} - \dfrac{1}{-(2x-3)}$

$$= \frac{2x}{2x-3} - \frac{-1}{2x-3}$$

$$= \frac{2x+1}{2x-3}$$

67. $a^2 - 9 = (a+3)(a-3)$

LCD: $(a+3)(a-3)$

$$\frac{1}{a-3} - \frac{1}{a+3} + \frac{2a}{a^2-9}$$

$$= \frac{1 \cdot (a+3)}{(a-3) \cdot (a+3)} - \frac{1 \cdot (a-3)}{(a+3) \cdot (a-3)} + \frac{2a}{(a+3)(a-3)}$$

$$= \frac{(a+3) - (a-3) + 2a}{(a+3)(a-3)}$$

$$= \frac{a+3-a+3+2a}{(a+3)(a-3)}$$

$$= \frac{2a+6}{(a+3)(a-3)}$$

$$= \frac{2\overset{1}{\cancel{(a+3)}}}{\cancel{(a+3)}(a-3)}$$

$$= \frac{2}{a-3}$$

69. $\dfrac{2x^2+3x}{x^2-2x-63} + \dfrac{7-x}{x^2-2x-63} - \dfrac{x^2-3x+21}{x^2-2x-63}$

$$= \frac{2x^2+3x+7-x-x^2+3x-21}{(x-9)(x+7)}$$

$$= \frac{x^2+5x-14}{(x-9)(x+7)}$$

$$= \frac{\overset{1}{\cancel{(x+7)}}(x-2)}{(x-9)\underset{1}{\cancel{(x+7)}}}$$

$$= \frac{x-2}{x-9}$$

71. Let x represent the first even integer.

Then $x + 2$ represents the next even integer.

$$\frac{1}{x} + \frac{1}{x+2} = \frac{1 \cdot (x+2)}{x \cdot (x+2)} + \frac{1 \cdot x}{(x+2) \cdot x}$$

$$= \frac{x+2}{x(x+2)} + \frac{x}{x(x+2)}$$

$$= \frac{x+2+x}{x(x+2)}$$

$$= \frac{2x+2}{x(x+2)}$$

73. The perimeter is two times the length plus two times the width.

$$\frac{2(2x+1)}{5} + \frac{2(4)}{3x+1}$$

$$= \frac{2(2x+1) \cdot (3x+1)}{5 \cdot (3x+1)} + \frac{2(4) \cdot 5}{(3x+1) \cdot 5}$$

$$= \frac{2(2x+1)(3x+1)}{5(3x+1)} + \frac{40}{5(3x+1)}$$

$$= \frac{2(6x^2+5x+1)}{5(3x+1)} + \frac{40}{5(3x+1)}$$

$$= \frac{12x^2+10x+2+40}{5(3x+1)}$$

$$= \frac{12x^2+10x+42}{5(3x+1)}$$

$$= \frac{2(6x^2+5x+21)}{5(3x+1)}$$

Exercises 5.4

1. $\dfrac{3}{7} \cdot \dfrac{14}{27} = \dfrac{3 \cdot 14}{7 \cdot 27} = \dfrac{2}{9}$

3. $\dfrac{x}{2} \cdot \dfrac{y}{6} = \dfrac{x \cdot y}{2 \cdot 6} = \dfrac{xy}{12}$

5. $\dfrac{3a}{2} \cdot \dfrac{4}{a^2} = \dfrac{3\overset{2}{\cancel{a}} \cdot \overset{2}{\cancel{4}}}{\underset{1}{\cancel{2}} \cdot \underset{a}{\cancel{a^2}}} = \dfrac{6}{a}$

7. $\dfrac{3x^3y}{10xy^3} \cdot \dfrac{5xy^2}{9xy^2} = \dfrac{3x^3y \cdot 5xy^2}{10xy^3 \cdot 9xy^2} = \dfrac{x^2}{6y^2}$

Divide by the common factors of 15, x^2, and y^3.

9. $\dfrac{-4ab^2}{15a^3} \cdot \dfrac{25ab}{-16b^3} = \dfrac{-4ab^2 \cdot 25ab}{15a^3 \cdot (-16b^3)} = \dfrac{5}{12a}$

Divide by the common factors of -20, a^2, and b^3.

11. $\dfrac{-3m^3n}{10mn^3} \cdot \dfrac{5mn^2}{-9mn^3} = \dfrac{-3m^3n \cdot 5mn^2}{10mn^3 \cdot (-9mn^3)} = \dfrac{m^2}{6n^3}$

Divide by the common factors of -15, m^2, and n^3.

13. $\dfrac{x^2+5x}{3x^2} \cdot \dfrac{10x}{5x+25} = \dfrac{x(x+5)}{3x^2} \cdot \dfrac{10x}{5(x+5)}$

$$= \dfrac{\overset{1}{\cancel{x}}(\overset{1}{\cancel{x+5}}) \cdot \overset{2}{\cancel{10}}\overset{1}{\cancel{x}}}{\underset{1}{\cancel{3}}\,x^{2} \cdot \underset{1}{\cancel{5}}(\underset{1}{\cancel{x+5}})}$$

$$= \dfrac{2}{3}$$

15. $\dfrac{p^2-8p}{4p} \cdot \dfrac{12p^2}{p^2-64} = \dfrac{p(p-8)}{4p} \cdot \dfrac{12p^2}{(p-8)(p+8)}$

$$= \dfrac{\overset{1}{\cancel{p}}(\overset{1}{\cancel{p-8}}) \cdot \overset{3}{\cancel{12}}\,p^2}{\underset{1}{\cancel{4}}\,\underset{1}{\cancel{p}} \cdot (\underset{1}{\cancel{p-8}})(p+8)}$$

$$= \dfrac{3p^2}{p+8}$$

17. $\dfrac{m^2-4m-21}{3m^2} \cdot \dfrac{m^2+7m}{m^2-49}$

$$= \dfrac{(m-7)(m+3)}{3m^2} \cdot \dfrac{m(m+7)}{(m+7)(m-7)}$$

$$= \dfrac{(\overset{1}{\cancel{m-7}})(m+3) \cdot \overset{1}{\cancel{m}}(\overset{1}{\cancel{m+7}})}{3\underset{m}{\cancel{m^2}} \cdot (\underset{1}{\cancel{m+7}})(\underset{1}{\cancel{m-7}})}$$

$$= \dfrac{m+3}{3m}$$

19. $\dfrac{4r^2-1}{2r^2-9r-5} \cdot \dfrac{3r^2-13r-10}{9r^2-4}$

$$= \dfrac{(2r-1)(2r+1)}{(2r+1)(r-5)} \cdot \dfrac{(3r+2)(r-5)}{(3r-2)(3r+2)}$$

$$= \dfrac{(2r-1)(\overset{1}{\cancel{2r+1}}) \cdot (\overset{1}{\cancel{3r+2}})(\overset{1}{\cancel{r-5}})}{(\underset{1}{\cancel{2r+1}})(\underset{1}{\cancel{r-5}}) \cdot (3r-2)(\underset{1}{\cancel{3r+2}})}$$

$$= \dfrac{2r-1}{3r-2}$$

21. $\dfrac{x^2-4y^2}{x^2-xy-6y^2} \cdot \dfrac{7x^2-21xy}{5x-10y}$

$$= \dfrac{(x-2y)(x+2y)}{(x-3y)(x+2y)} \cdot \dfrac{7x(x-3y)}{5(x-2y)}$$

$$= \dfrac{(\overset{1}{\cancel{x-2y}})(\overset{1}{\cancel{x+2y}}) \cdot 7x(\overset{1}{\cancel{x-3y}})}{(\underset{1}{\cancel{x-3y}})(\underset{1}{\cancel{x+2y}}) \cdot 5(\underset{1}{\cancel{x-2y}})}$$

$$= \dfrac{7x}{5}$$

23. $\dfrac{2x-6}{x^2+2x} \cdot \dfrac{3x}{3-x} = \dfrac{2(x-3)}{x(x+2)} \cdot \dfrac{3x}{-1(x-3)}$

$$= \dfrac{2(\overset{1}{\cancel{x-3}}) \cdot 3\overset{1}{\cancel{x}}}{\underset{1}{\cancel{x}}(x+2) \cdot (-1)(\underset{1}{\cancel{x-3}})}$$

$$= \dfrac{-6}{x+2}$$

25. $\dfrac{5}{8} \div \dfrac{15}{16} = \dfrac{5}{8} \cdot \dfrac{16}{15} = \dfrac{\overset{1}{\cancel{5}} \cdot \overset{2}{\cancel{16}}}{\underset{1}{\cancel{8}} \cdot \underset{3}{\cancel{15}}} = \dfrac{2}{3}$

27. $\dfrac{5}{x^2} \div \dfrac{10}{x} = \dfrac{5}{x^2} \cdot \dfrac{x}{10} = \dfrac{\overset{1}{\cancel{5}}\overset{1}{\cancel{x}}}{\underset{2}{\cancel{10}}\,\underset{x}{\cancel{x^2}}} = \dfrac{1}{2x}$

29. $\dfrac{4x^2y^2}{9x^3} \div \dfrac{8y^2}{27xy} = \dfrac{4x^2y^2}{9x^3} \cdot \dfrac{27xy}{8y^2} = \dfrac{3y}{2}$

Divide by the common factors of 36, x^3, and y^2.

31. $\dfrac{3x+6}{8} \div \dfrac{5x+10}{6} = \dfrac{3x+6}{8} \cdot \dfrac{6}{5x+10}$

$$= \dfrac{3(\overset{1}{\cancel{x+2}}) \cdot \overset{3}{\cancel{6}}}{\underset{4}{\cancel{8}} \cdot 5(\underset{1}{\cancel{x+2}})} = \dfrac{9}{20}$$

33. $\dfrac{4a-12}{5a+15} \div \dfrac{8a^2}{a^2+3a} = \dfrac{4a-12}{5a+15} \cdot \dfrac{a^2+3a}{8a^2}$

$$= \dfrac{\overset{1}{\cancel{4}}(a-3) \cdot \overset{1}{\cancel{a}}(\overset{1}{\cancel{a+3}})}{5(\underset{1}{\cancel{a+3}}) \cdot \underset{2}{\cancel{8}}\,\underset{a}{\cancel{a^2}}}$$

$$= \dfrac{a-3}{10a}$$

35. $\dfrac{x^2+2x-8}{9x^2} \div \dfrac{x^2-16}{3x-12} = \dfrac{x^2+2x-8}{9x^2} \cdot \dfrac{3x-12}{x^2-16}$

$$= \dfrac{(x+4)(x-2) \cdot \overset{1}{\cancel{3}}(\overset{1}{\cancel{x-4}})}{\underset{3}{\cancel{9}}\,x^2 \cdot (\underset{1}{\cancel{x-4}})(\underset{1}{\cancel{x+4}})}$$

$$= \dfrac{x-2}{3x^2}$$

37.

$$\frac{x^2-9}{2x^2-6x} \div \frac{2x^2+5x-3}{4x^2-1}$$

$$= \frac{x^2-9}{2x^2-6x} \cdot \frac{4x^2-1}{2x^2+5x-3}$$

$$= \frac{(x+3)(x-3) \cdot (2x-1)(2x+1)}{2x(x-3) \cdot (2x-1)(x+3)}$$

$$= \frac{2x+1}{2x}$$

39.

$$\frac{a^2-9b^2}{4a^2+12ab} \div \frac{a^2-ab-6b^2}{12ab}$$

$$= \frac{a^2-9b^2}{4a^2+12ab} \cdot \frac{12ab}{a^2-ab-6b^2}$$

$$= \frac{(a-3b)(a+3b) \cdot 12\,a\,b}{4\,a(a+3b) \cdot (a-3b)(a+2b)}$$

$$= \frac{3b}{a+2b}$$

41.

$$\frac{x^2-16y^2}{3x^2-12xy} \div (x^2+4xy)$$

$$= \frac{x^2-16y^2}{3x^2-12xy} \cdot \frac{1}{x^2+4xy}$$

$$= \frac{(x-4y)(x+4y) \cdot 1}{3x(x-4y) \cdot x(x+4y)}$$

$$= \frac{1}{3x^2}$$

43.

$$\frac{x-7}{2x+6} \div \frac{21-3x}{x^2+3x} = \frac{x-7}{2x+6} \cdot \frac{x^2+3x}{21-3x}$$

$$= \frac{(x-7) \cdot x(x+3)}{2(x+3)(-3)(x-7)}$$

$$= \frac{-x}{6}$$

45.

$$\frac{x^2-2x-8}{2x-8} \cdot \frac{x^2+5x}{x^2+5x+6} \div \frac{x^2+2x-15}{x^2-9}$$

$$= \frac{x^2-2x-8}{2x-8} \cdot \frac{x^2+5x}{x^2+5x+6} \cdot \frac{x^2-9}{x^2+2x-15}$$

$$= \frac{(x-4)(x+2) \cdot x(x+5) \cdot (x-3)(x+3)}{2(x-4) \cdot (x+3)(x+2) \cdot (x+5)(x-3)} = \frac{x}{2}$$

47.

$$\frac{x^2+5x}{3x-6} \cdot \frac{x^2-4}{3x^2+15x} \cdot \frac{6x}{x^2+6x+8}$$

$$= \frac{x(x+5) \cdot (x+2)(x-2) \cdot 6\,x}{3(x-2) \cdot 3x(x+5) \cdot (x+4)(x+2)}$$

$$= \frac{2x}{3(x+4)}$$

49.

$$\frac{\frac{2}{3}}{\frac{6}{8}} = \frac{\frac{2}{3} \cdot 24}{\frac{6}{8} \cdot 24} = \frac{2 \cdot 8}{6 \cdot 3} = \frac{8}{9}$$

51.

$$\frac{1+\frac{1}{2}}{2+\frac{1}{4}} = \frac{\left(1+\frac{1}{2}\right) \cdot 4}{\left(2+\frac{1}{4}\right) \cdot 4} = \frac{4+2}{8+1} = \frac{6}{9} = \frac{2}{3}$$

53.

$$\frac{\frac{x}{8}}{\frac{x^2}{4}} = \frac{\frac{x}{8} \cdot 8}{\frac{x^2}{4} \cdot 8} = \frac{x}{2x^2} = \frac{1}{2x}$$

55.

$$\frac{\frac{3}{a}}{\frac{2}{a^2}} = \frac{\frac{3}{a} \cdot a^2}{\frac{2}{a^2} \cdot a^2} = \frac{3a}{2}$$

57.

$$\frac{\frac{y+1}{y}}{\frac{y-1}{2y}} = \frac{y+1}{y} \cdot \frac{2y}{y-1} = \frac{2(y+1)}{y-1}$$

59.

$$\frac{2-\frac{1}{x}}{2+\frac{1}{x}} = \frac{\left(2-\frac{1}{x}\right) \cdot x}{\left(2+\frac{1}{x}\right) \cdot x} = \frac{2x-1}{2x+1}$$

61.

$$\frac{3-\frac{x}{y}}{\frac{6}{y}} = \frac{\left(3-\frac{x}{y}\right)}{\left(\frac{6}{y}\right) \cdot y} = \frac{3y-x}{6}$$

63.

$$\frac{\frac{x^2}{y^2}-1}{\frac{x}{y}+1} = \frac{\left(\frac{x^2}{y^2}-1\right) \cdot y^2}{\left(\frac{x}{y}+1\right) \cdot y^2}$$

$$= \frac{x^2-y^2}{xy+y^2}$$

$$= \frac{(x+y)(x-y)}{y(x+y)}$$

$$= \frac{x-y}{y}$$

65.
$$\frac{1+\frac{3}{x}-\frac{4}{x^2}}{1+\frac{2}{x}-\frac{3}{x^2}} = \frac{\left(1+\frac{3}{x}-\frac{4}{x^2}\right)\cdot x^2}{\left(1+\frac{2}{x}-\frac{3}{x^2}\right)\cdot x^2}$$
$$= \frac{x^2+3x-4}{x^2+2x-3}$$
$$= \frac{(x+4)(x-1)}{(x+3)(x-1)}$$
$$= \frac{x+4}{x+3}$$

67.
$$\frac{\frac{2}{x}-\frac{1}{xy}}{\frac{1}{xy}+\frac{2}{y}} = \frac{\left(\frac{2}{x}-\frac{1}{xy}\right)\cdot xy}{\left(\frac{1}{xy}+\frac{2}{y}\right)\cdot xy} = \frac{2y-1}{1+2x}$$

69.
$$\frac{\frac{2}{x-1}+1}{1-\frac{3}{x-1}} = \frac{\left(\frac{2}{x-1}+1\right)\cdot(x-1)}{\left(1-\frac{3}{x-1}\right)\cdot(x-1)} = \frac{2+(x-1)}{(x-1)-3} = \frac{x+1}{x-4}$$

71.
$$\frac{1-\frac{1}{y-1}}{y-\frac{8}{y+2}} = \frac{\left(1-\frac{1}{y-1}\right)\cdot(y-1)(y+2)}{\left(y-\frac{8}{y+2}\right)\cdot(y-1)(y+2)}$$
$$= \frac{(y-1)(y+2)-(y+2)}{y(y-1)(y+2)-8(y-1)}$$
$$= \frac{(y+2)[(y-1)-1]}{(y-1)[y(y+2)-8]}$$
$$= \frac{(y+2)(y-2)}{(y-1)(y^2+2y-8)}$$
$$= \frac{(y+2)(y-2)}{(y-1)(y+4)(y-2)}$$
$$= \frac{y+2}{(y-1)(y+4)}$$

73.
$$1+\frac{1}{1+\frac{1}{x}} = 1+\frac{1\cdot x}{\left(1+\frac{1}{x}\right)\cdot x}$$
$$= 1+\frac{x}{x+1}$$
$$= \frac{x+1}{x+1}+\frac{x}{x+1}$$
$$= \frac{x+1+x}{x+1}$$
$$= \frac{2x+1}{x+1}$$

75.
$$\frac{\frac{2}{3}}{\frac{1}{4}} = \frac{2}{3}\cdot\frac{4}{1} = \frac{8}{3}$$

77.
$$\frac{\frac{7}{10}}{\frac{4}{5}} = \frac{7}{10}\cdot\frac{5}{4} = \frac{7}{8}$$

79. The area of a rectangle is the product of the two sides.
$$\frac{2x+6}{12x-15}\cdot\frac{4x-5}{x+3} = \frac{2(x+3)\cdot(4x-5)}{3(4x-5)\cdot(x+3)} = \frac{2}{3}$$

81. Writing exercise

Exercises 5.5

1. $\frac{x}{15}$; None

3. $\frac{17}{x}$; $x=0$ is the excluded value.

5. $\frac{3}{x-2}$; If $x-2=0$, then $x=2$. So, 2 is the excluded value.

7. $\frac{-5}{x+4}$; If $x+4=0$, then $x=-4$. So, -4 is the excluded value.

9. $\frac{x-5}{2}$; None

11. $\frac{3x}{(x+1)(x-2)}$; If $(x+1)(x-2)=0$, then $x+1=0$ or $x-2=0$. So, $x=-1$ and $x=2$ are the excluded values.

13. $\frac{x-1}{(2x-1)(x+3)}$; If $(2x-1)(x+3)=0$, then $2x-1=0$ or $x+3=0$. So, $x=\frac{1}{2}$ and $x=-3$ are the excluded values.

15. $\frac{7}{x^2-9} = \frac{7}{(x+3)(x-3)}$; If $(x+3)(x-3)=0$, then $x=-3$ or $x=3$. So, -3 and 3 are the excluded values.

17. $\frac{x+3}{x^2-7x+12} = \frac{x+3}{(x-4)(x-3)}$;
If $(x-4)(x-3)=0$, then $x-4=0$ or $x-3=0$. So, 4 and 3 are the excluded values.

19. $\dfrac{2x-1}{3x^2+x-2}=\dfrac{2x-1}{(3x-2)(x+1)}$;

If $(3x-2)(x+1)=0$, then $x=\dfrac{2}{3}$ or $x=-1$.

So, $\dfrac{2}{3}$ and -1 are the excluded values.

21. $\dfrac{x}{2}+3=6$

The LCD is 2.

$$2\cdot\dfrac{x}{2}+2\cdot 3=2\cdot 6$$
$$x+6=12$$
$$x=6$$

23. $\dfrac{x}{2}-\dfrac{x}{3}=2$

The LCD is 6.

$$6\cdot\dfrac{x}{2}-6\cdot\dfrac{x}{3}=6\cdot 2$$
$$3x-2x=12$$
$$x=12$$

25. $\dfrac{x}{5}-\dfrac{1}{3}=\dfrac{x-7}{3}$

The LCD is 15.

$$15\cdot\dfrac{x}{5}-15\cdot\dfrac{1}{3}=15\cdot\dfrac{x-7}{3}$$
$$3x-5=5(x-7)$$
$$3x-5=5x-35$$
$$30=2x$$
$$15=x \text{ or } x=15$$

27. $\dfrac{x}{4}-\dfrac{1}{5}=\dfrac{4x+3}{20}$

The LCD is 20.

$$20\cdot\dfrac{x}{4}-20\cdot\dfrac{1}{5}=20\cdot\dfrac{4x+3}{20}$$
$$5x-4=4x+3$$
$$x=7$$

29. $\dfrac{3}{x}+2=\dfrac{7}{x}$

The LCD is x.

$$x\cdot\dfrac{3}{x}+x\cdot 2=x\cdot\dfrac{7}{x}$$
$$3+2x=7$$
$$2x=4$$
$$x=2$$

31. $\dfrac{4}{x}+\dfrac{3}{4}=\dfrac{10}{x}$

The LCD is $4x$.

$$4x\cdot\dfrac{4}{x}+4x\cdot\dfrac{3}{4}=4x\cdot\dfrac{10}{x}$$
$$16+3x=40$$
$$3x=24$$
$$x=8$$

33. $\dfrac{5}{2x}-\dfrac{1}{x}=\dfrac{9}{2x^2}$

The LCD is $2x^2$.

$$2x^2\cdot\dfrac{5}{2x}-2x^2\cdot\dfrac{1}{x}=2x^2\cdot\dfrac{9}{2x^2}$$
$$5x-2x=9$$
$$3x=9$$
$$x=3$$

35. $\dfrac{2}{x-3}+1=\dfrac{7}{x-3}$

The LCD is $x-3$.

$$(x-3)\cdot\dfrac{2}{x-3}+(x-3)\cdot 1=(x-3)\cdot\dfrac{7}{x-3}$$
$$2+(x-3)=7$$
$$x-1=7$$
$$x=8$$

37. $\dfrac{12}{x+3}=\dfrac{x}{x+3}+2$

The LCD is $x+3$.

$$(x+3)\cdot\dfrac{12}{x+3}=(x+3)\cdot\dfrac{x}{x+3}+(x+3)\cdot 2$$
$$12=x+2(x+3)$$
$$12=x+2x+6$$
$$6=3x$$
$$2=x \text{ or } x=2$$

39. $\dfrac{3}{x-5}+4=\dfrac{2x+5}{x-5}$

The LCD is $x-5$.

$$(x-5)\cdot\dfrac{3}{x-5}+(x-5)\cdot 4=(x-5)\cdot\dfrac{2x+5}{x-5}$$
$$3+4(x-5)=2x+5$$
$$3+4x-20=2x+5$$
$$2x=22$$
$$x=11$$

41. $\dfrac{2}{x+3} + \dfrac{1}{2} = \dfrac{x+6}{x+3}$

The LCD is $2(x+3)$.

$$2(x+3) \cdot \dfrac{2}{x+3} + 2(x+3) \cdot \dfrac{1}{2} = 2(x+3) \cdot \dfrac{x+6}{x+3}$$
$$4 + (x+3) = 2(x+6)$$
$$x + 7 = 2x + 12$$
$$-5 = x \text{ or } x = -5$$

43. $\dfrac{x}{3x+12} + \dfrac{x-1}{x+4} = \dfrac{5}{3}$

$$\dfrac{x}{3(x+4)} + \dfrac{x-1}{x+4} = \dfrac{5}{3}$$

The LCD is $3(x+4)$.

$$3(x+4) \cdot \dfrac{x}{3(x+4)} + 3(x+4) \cdot \dfrac{x-1}{x+4} = 3(x+4) \cdot \dfrac{5}{3}$$
$$x + 3(x-1) = 5(x+4)$$
$$x + 3x - 3 = 5x + 20$$
$$-23 = x \text{ or } x = -23$$

45. $\dfrac{x}{x-3} - 2 = \dfrac{3}{x-3}$

The LCD is $x-3$.

$$(x-3) \cdot \dfrac{x}{x-3} - 2(x-3) = (x-3) \cdot \dfrac{3}{x-3}$$
$$x - 2(x-3) = 3$$
$$x - 2x + 6 = 3$$
$$3 = x \text{ or } x = 3$$

But, $x = 3$ makes the fraction in the original equation undefined. Since 3 is an excluded value, the equation has no solution.

47. $\dfrac{x-1}{x+3} - \dfrac{x-3}{x} = \dfrac{3}{x^2 + 3x}$

$$\dfrac{x-1}{x+3} - \dfrac{x-3}{x} = \dfrac{3}{x(x+3)}$$

The LCD is $x(x+3)$.

$$x(x+3) \cdot \dfrac{x-1}{x+3} - x(x+3) \cdot \dfrac{x-3}{x} = x(x+3) \cdot \dfrac{3}{x(x+3)}$$
$$x(x-1) - (x-3)(x+3) = 3$$
$$x^2 - x - (x^2 - 9) = 3$$
$$-x + 9 = 3$$
$$6 = x \text{ or } x = 6$$

49. $\dfrac{1}{x-2} - \dfrac{2}{x+2} = \dfrac{2}{x^2-4}$

$\dfrac{1}{x-2} - \dfrac{2}{x+2} = \dfrac{2}{(x+2)(x-2)}$

The LCD is $(x+2)(x-2)$.

$(x+2)(x-2) \cdot \dfrac{1}{x-2} - (x+2)(x-2) \cdot \dfrac{2}{x+2} = (x+2)(x-2) \cdot \dfrac{2}{(x+2)(x-2)}$

$$(x+2) - 2(x-2) = 2$$
$$x + 2 - 2x + 4 = 2$$
$$6 - x = 2$$
$$4 = x \text{ or } x = 4$$

51. $\dfrac{5}{x-4} = \dfrac{1}{x+2} - \dfrac{2}{x^2-2x-8}$

$\dfrac{5}{x-4} = \dfrac{1}{x+2} - \dfrac{2}{(x-4)(x+2)}$

The LCD is $(x-4)(x+2)$.

$(x-4)(x+2) \cdot \dfrac{5}{x-4} = (x-4)(x+2) \cdot \dfrac{1}{x+2} - (x-4)(x+2) \cdot \dfrac{2}{(x-4)(x+2)}$

$$5(x+2) = (x-4) - 2$$
$$5x + 10 = x - 6$$
$$4x = -16$$
$$x = -4$$

53. $\dfrac{3}{x-1} - \dfrac{1}{x+9} = \dfrac{18}{x^2+8x-9}$

$\dfrac{3}{x-1} - \dfrac{1}{x+9} = \dfrac{18}{(x+9)(x-1)}$

The LCD is $(x+9)(x-1)$.

$(x+9)(x-1) \cdot \dfrac{3}{x-1} - (x+9)(x-1) \cdot \dfrac{1}{x+9} = (x+9)(x-1) \cdot \dfrac{18}{(x+9)(x-1)}$

$$3(x+9) - (x-1) = 18$$
$$3x + 27 - x + 1 = 18$$
$$2x = -10$$
$$x = -5$$

55. $\dfrac{3}{x+3} + \dfrac{25}{x^2+x-6} = \dfrac{5}{x-2}$

$\dfrac{3}{x+3} + \dfrac{25}{(x+3)(x-2)} = \dfrac{5}{x-2}$

The LCD is $(x+3)(x-2)$.

$(x+3)(x-2) \cdot \dfrac{3}{x+3} + (x+3)(x-2) \cdot \dfrac{25}{(x+3)(x-2)} = (x+3)(x-2) \cdot \dfrac{5}{x-2}$

$$3(x-2) + 25 = 5(x+3)$$
$$3x - 6 + 25 = 5x + 15$$
$$4 = 2x$$
$$2 = x \text{ or } x = 2$$

But, $x = 2$ makes a fraction in the original equation undefined. Since 2 is an excluded value, the equation has no solution.

57.
$$\frac{7}{x-5} - \frac{3}{x+5} = \frac{40}{x^2-25}$$
$$\frac{7}{x-5} - \frac{3}{x+5} = \frac{40}{(x+5)(x-5)}$$
The LCD is $(x+5)(x-5)$.
$$(x+5)(x-5)\cdot\frac{7}{x-5} - (x+5)(x-5)\cdot\frac{3}{x+5} = (x+5)(x-5)\cdot\frac{40}{(x+5)(x-5)}$$
$$7(x+5) - 3(x-5) = 40$$
$$7x+35-3x+15 = 40$$
$$4x = -10$$
$$x = -\frac{10}{4} = -\frac{5}{2}$$

59.
$$\frac{2x}{x-3} + \frac{2}{x-5} = \frac{3x}{x^2-8x+15}$$
$$\frac{2x}{x-3} + \frac{2}{x-5} = \frac{3x}{(x-3)(x-5)}$$
The LCD is $(x-3)(x-5)$.
$$(x-3)(x-5)\cdot\frac{2x}{x-3} + (x-3)(x-5)\cdot\frac{2}{x-5} = (x-3)(x-5)\cdot\frac{3x}{(x-3)(x-5)}$$
$$2x(x-5) + 2(x-3) = 3x$$
$$2x^2 - 10x + 2x - 6 = 3x$$
$$2x^2 - 11x - 6 = 0$$
$$(2x+1)(x-6) = 0$$
$$2x+1=0 \quad \text{or } x-6=0$$
$$x=-\frac{1}{2} \text{ or } \quad x=6$$

61.
$$\frac{2x}{x+2} = \frac{5}{x^2-x-6} - \frac{1}{x-3}$$
$$\frac{2x}{x+2} = \frac{5}{(x+2)(x-3)} - \frac{1}{x-3}$$
The LCD is $(x+2)(x-3)$.
$$(x+2)(x-3)\cdot\frac{2x}{x+2} = (x+2)(x-3)\cdot\frac{5}{(x+2)(x-3)} - (x+2)(x-3)\cdot\frac{1}{x-3}$$
$$2x(x-3) = 5 - (x+2)$$
$$2x^2 - 6x = 5 - x - 2$$
$$2x^2 - 5x - 3 = 0$$
$$(2x+1)(x-3) = 0$$
$$2x+1=0 \quad \text{or } x-3=0$$
$$x=-\frac{1}{2} \text{ or } \quad x=3$$

But since 3 is an excluded value, the solution is $-\frac{1}{2}$.

63. $\dfrac{7}{x-2} + \dfrac{16}{x+3} = 3$

The LCD is $(x-2)(x+3)$.

$$(x-2)(x+3) \cdot \dfrac{7}{x-2} + (x-2)(x+3) \cdot \dfrac{16}{x+3} = 3(x-2)(x+3)$$
$$7(x+3) + 16(x-2) = 3(x-2)(x+3)$$
$$7x + 21 + 16x - 32 = 3x^2 + 3x - 18$$
$$3x^2 - 20x - 7 = 0$$
$$(3x+1)(x-7) = 0$$

$3x + 1 = 0$ or $x - 7 = 0$

$x = -\dfrac{1}{3}$ or $x = 7$

65. $\dfrac{11}{x-3} - 1 = \dfrac{10}{x+3}$

The LCD is $(x-3)(x+3)$.

$$(x-3)(x+3) \cdot \dfrac{11}{x-3} - 1(x-3)(x+3) = (x-3)(x+3) \cdot \dfrac{10}{x+3}$$
$$11(x+3) - (x-3)(x+3) = 10(x-3)$$
$$11x + 33 - x^2 + 9 = 10x - 30$$
$$x^2 - x - 72 = 0$$
$$(x-9)(x+8) = 0$$

$x - 9 = 0$ or $x + 8 = 0$

$x = 9$ or $x = -8$

67. $\dfrac{x}{11} = \dfrac{12}{33}$

$33x = 132$

$x = 4$

69. $\dfrac{5}{8} = \dfrac{20}{x}$

$5x = 160$

$x = 32$

71. $\dfrac{x+1}{5} = \dfrac{20}{25}$

$25(x+1) = 100$

$25x + 25 = 100$

$25x = 75$

$x = 3$

73. $\dfrac{3}{5} = \dfrac{x-1}{20}$

$60 = 5(x-1)$

$60 = 5x - 5$

$65 = 5x$

$13 = x$ or $x = 13$

75.
$$\frac{x}{6} = \frac{x+5}{16}$$
$$16x = 6(x+5)$$
$$16x = 6x + 30$$
$$10x = 30$$
$$x = 3$$

77.
$$\frac{x}{x+7} = \frac{10}{17}$$
$$17x = 10(x+7)$$
$$17x = 10x + 70$$
$$7x = 70$$
$$x = 10$$

79.
$$\frac{2}{x-1} = \frac{6}{x+9}$$
$$2(x+9) = 6(x-1)$$
$$2x + 18 = 6x - 6$$
$$24 = 4x$$
$$6 = x \text{ or } x = 6$$

81.
$$\frac{1}{x+3} = \frac{7}{x^2 - 9}$$
$$\frac{1}{x+3} = \frac{7}{(x-3)(x+3)}$$
The LCD is $(x-3)(x+3)$.
$$(x-3)(x+3) \cdot \frac{1}{x+3} = (x-3)(x+3) \cdot \frac{7}{(x-3)(x+3)}$$
$$x - 3 = 7$$
$$x = 10$$

Exercises 5.6

1. Let x be the unknown number.
$$\frac{2}{3}x + \frac{1}{2}x = 35$$
$$4x + 3x = 210$$
$$7x = 210$$
$$x = 30$$
The number is 30.

3. Let x be the unknown number.
$$\frac{2}{5}x - \frac{1}{4}x = 3$$
$$8x - 5x = 60$$
$$3x = 60$$
$$x = 20$$
The number is 20.

5. Let x be the first integer.
Then $x + 1$ is the next consecutive integer.

$$\frac{1}{3}x + \frac{1}{2}(x+1) = 13$$
$$2x + 3(x+1) = 78$$
$$2x + 3x + 3 = 78$$
$$5x = 75$$
$$x = 15; \ x + 1 = 16$$
The integers are 15, 16.

7. Let x be one number.
Then $2x$ is the other number.
$$\frac{1}{x} + \frac{1}{2x} = \frac{1}{4}$$
$$4 + 2 = x$$
$$x = 6; \ 2x = 12$$
The numbers are 6, 12.

9. Let x be one number.
Then $4x$ is the other number.
$$\frac{1}{x} + \frac{1}{4x} = \frac{5}{12}$$
$$12 + 3 = 5x$$
$$15 = 5x$$
$$x = 3; \ 4x = 12$$
The numbers are 3, 12.

11. Let x be one number.
Then $5x$ is the other number.
$$\frac{1}{x} + \frac{1}{5x} = \frac{6}{35}$$
$$35 + 7 = 6x$$
$$42 = 6x$$
$$x = 7; \ 5x = 35$$
The numbers are 7, 35.

13. Let x be the number.
$$\frac{1}{x} - \frac{1}{5x} = \frac{4}{25}$$
$$25 - 5 = 4x$$
$$20 = 4x$$
$$x = 5$$
The number is 5.

15. Let r be his rate bicycling. Then $r + 30$ is his rate driving.

	Distance	Rate	Time
Bicycling	50	r	$\dfrac{50}{r}$
Driving	125	$r + 30$	$\dfrac{125}{r+30}$

Time bicycling = Time driving
$$\frac{50}{r} = \frac{125}{r+30}$$
$$50(r+30) = 125r$$
$$50r + 1500 = 125r$$
$$1500 = 75r$$
$$r = 20$$
$$r + 30 = 50$$
His bicycling rate is 20 mi/h and his driving rate is 50 mi/h.

17. Let r be the rate of the express bus. Then $r - 10$ is the rate of the local bus.

	Distance	Rate	Time
Express	275	r	$\dfrac{275}{r}$
Local	225	$r-10$	$\dfrac{225}{r-10}$

Time for express bus = Time for local bus
$$\frac{275}{r} = \frac{225}{r-10}$$
$$275(r-10) = 225r$$
$$275r - 2750 = 225r$$
$$50r = 2750$$
$$r = 55$$
$$r - 10 = 45$$
The rate of the express bus is 55 mi/h, and the rate of the local bus is 45 mi/h.

19. Let r be the speed of the freight train. Then $r + 25$ is the speed of the passenger train.

	Distance	Rate	Time
Passenger	325	$r+25$	$\dfrac{325}{r+25}$
Freight	200	r	$\dfrac{200}{r}$

Passenger train time = Freight train time
$$\frac{325}{r+25} = \frac{200}{r}$$
$$325r = 200(r+25)$$
$$325r = 200r + 5000$$
$$125r = 5000$$
$$r = 40$$
$$r + 25 = 65$$
The frieght train's speed is 40 mi/h and the passenger train's speed is 65 mi/h.

21. Let t be the time he took on the second day. Then $t + 2$ is the time he took on the first day.

	Distance	Rate	Time
First day	240	$\dfrac{240}{t+2}$	$t+2$
Second Day	144	$\dfrac{144}{t}$	t

First day rate = Second day rate
$$\frac{240}{t+2} = \frac{144}{t}$$
$$240t = 144(t+2)$$
$$240t = 144t + 288$$
$$96t = 288$$
$$t = 3$$
$$t + 2 = 5$$
His driving time was 5 h for the first day and 3 h for the second day.

23. Let t be the time of the 480-mile flight. Then $t + 3$ is the time of the 1200-mile flight.

	Distance	Rate	Time
Trip 1	480	$\dfrac{480}{t}$	t
Trip 2	1200	$\dfrac{1200}{t+3}$	$t+3$

Trip 1 rate = Trip 2 rate
$$\frac{480}{t} = \frac{1200}{t+3}$$
$$480(t+3) = 1200t$$
$$480t + 1440 = 1200t$$
$$1440 = 720t$$
$$t = 2$$
$$t + 3 = 5$$
The 480-mile flight took 2 h and the 1200-mile flight took 5 h.

25. Let r be their rate in still water. Then $r - 2$ is their rate against the current and $r + 2$ is their rate with the current.

	Distance	Rate	Time
Upstream	6	$r - 2$	$\dfrac{6}{r - 2}$
Downstream	6	$r + 2$	$\dfrac{6}{r + 2}$

Time upstream + Time downstream = 4

$$\frac{6}{r - 2} + \frac{6}{r + 2} = 4$$
$$6(r + 2) + 6(r - 2) = 4(r + 2)(r - 2)$$
$$6r + 12 + 6r - 12 = 4(r^2 - 4)$$
$$12r = 4r^2 - 16$$
$$r^2 - 3r - 4 = 0$$
$$(r - 4)(r + 1) = 0$$
$$r - 4 = 0 \text{ or } r + 1 = 0$$
$$r = 4 \text{ or } \quad r = -1$$

Reject $r = -1$ since rate is positive.
They can paddle 4 mi/h in still water.

27. Let x be the amount of pure alcohol to be added.

	Amount	Strength	Total
Pure Alcohol	x	1.00	$x(1.00)$
25% Alcohol	40	0.25	$40(0.25)$

$$1.00x + 0.25(40) = 0.40(40 + x)$$
$$100x + 1000 = 40(40 + x)$$
$$100x + 1000 = 1600 + 40x$$
$$60x = 600$$
$$x = 10$$

The amount of pure alcohol to be added is 10 oz.

29. Let x be the speed of the plane in ft/s.
$$\frac{60}{88} = \frac{150}{x}$$
$$60x = 13,200$$
$$x = 220$$
The plane's speed is 220 ft/s.

31. Let x be the amount of gasoline.
$$\frac{5}{160} = \frac{x}{384}$$
$$1920 = 160x$$
$$12 = x$$
The 384-mi trip will take 12 gal of gasoline.

33. Let x be the amount Sveta earns in 1 year.
$$\frac{13,500}{20} = \frac{x}{52}$$
$$702,000 = 20x$$
$$35,100 = x$$
She will earn \$35,100 in 1 year.

35. Let x be the number of ladybug beetles. Then $110 - x$ is the number of praying mantises.
$$\frac{x}{110 - x} = \frac{7}{4}$$
$$4x = 7(110 - x)$$
$$4x = 770 - 7x$$
$$11x = 770$$
$$x = 70; \ 110 - x = 40$$
There are 70 ladybugs and 40 praying mantises in a package.

37. Let x be the amount the brother receives. Then $12,000x$ is the amount the sister recieves.
$$\frac{x}{12,000 - x} = \frac{2}{3}$$
$$3x = 2(12,000 - x)$$
$$3x = 24,000 - 2x$$
$$5x = 24,000$$
$$x = 4800; \ 12,000 - x = 7200$$
The brother receives \$4800 and sister receives \$7200.

Summary Exercises for Chapter 5

1. $\dfrac{6a^2}{9a^3} = \dfrac{2 \cdot 3 \cdot a \cdot a}{3 \cdot 3 \cdot a \cdot a \cdot a} = \dfrac{2}{3a}$

3. $\dfrac{w^2 - 25}{2w - 8} = \dfrac{(w - 5)(w + 5)}{2(w - 4)}$

$\dfrac{w^2 - 25}{2w - 8}$ cannot be reduced.

5.
$$\frac{m^2 - 2m - 3}{9 - m^2} = \frac{(m-3)(m+1)}{(3-m)(3+m)}$$
$$= \frac{-(m+1)}{3+m}$$
$$= \frac{-m-1}{m+3}$$

7.
$$\frac{x}{9} + \frac{2x}{9} = \frac{x + 2x}{9} = \frac{3x}{9} = \frac{x}{3}$$

9.
$$\frac{8}{x+2} + \frac{3}{x+2} = \frac{8+3}{x+2} = \frac{11}{x+2}$$

11.
$$\frac{7r - 3s}{4r} + \frac{r - s}{4r} = \frac{(7r - 3s) + (r - s)}{4r}$$
$$= \frac{7r - 3s + r - s}{4r}$$
$$= \frac{8r - 4s}{4r}$$
$$= \frac{4(2r - s)}{4r}$$
$$= \frac{2r - s}{r}$$

13.
$$\frac{5w - 6}{w - 4} - \frac{3w + 2}{w - 4} = \frac{(5w - 6) - (3w + 2)}{w - 4}$$
$$= \frac{5w - 6 - 3w - 2}{w - 4}$$
$$= \frac{2w - 8}{w - 4}$$
$$= \frac{2(w - 4)}{w - 4}$$
$$= 2$$

15.
$$\frac{5x}{6} + \frac{x}{3} = \frac{5x}{6} + \frac{x \cdot 2}{3 \cdot 2}$$
$$= \frac{5x}{6} + \frac{2x}{6}$$
$$= \frac{5x + 2x}{6}$$
$$= \frac{7x}{6}$$

17.
$$\frac{5}{2m} - \frac{3}{m^2} = \frac{5 \cdot m}{2m \cdot m} - \frac{3 \cdot 2}{m^2 \cdot 2}$$
$$= \frac{5m}{2m^2} - \frac{6}{2m^2}$$
$$= \frac{5m - 6}{2m^2}$$

19.
$$\frac{4}{x - 3} - \frac{1}{x} = \frac{4 \cdot x}{(x - 3) \cdot x} - \frac{1 \cdot (x - 3)}{x \cdot (x - 3)}$$
$$= \frac{4x}{x(x - 3)} - \frac{(x - 3)}{x(x - 3)}$$
$$= \frac{4x - (x - 3)}{x(x - 3)}$$
$$= \frac{4x - x + 3}{x(x - 3)}$$
$$= \frac{3x + 3}{x(x - 3)}$$

21.
$$\frac{5}{w - 5} - \frac{2}{w - 3}$$
$$= \frac{5 \cdot (w - 3)}{(w - 5) \cdot (w - 3)} - \frac{2 \cdot (w - 5)}{(w - 3)(w - 5)}$$
$$= \frac{5(w - 3) - 2(w - 5)}{(w - 5)(w - 3)}$$
$$= \frac{5w - 15 - 2w + 10}{(w - 5)(w - 3)}$$
$$= \frac{3w - 5}{(w - 5)(w - 3)}$$

23.
$$\frac{2}{3x - 3} - \frac{5}{2x - 2} = \frac{2}{3(x - 1)} - \frac{5}{2(x - 1)}$$
$$= \frac{2 \cdot 2}{3(x - 1) \cdot 2} - \frac{5 \cdot 3}{2(x - 1) \cdot 3}$$
$$= \frac{4}{6(x - 1)} - \frac{15}{6(x - 1)}$$
$$= \frac{4 - 15}{6(x - 1)}$$
$$= \frac{-11}{6(x - 1)}$$

25.
$$\frac{3a}{a^2 + 5a + 4} + \frac{2a}{a^2 - 1}$$

$$= \frac{3a}{(a+4)(a+1)} + \frac{2a}{(a+1)(a-1)}$$

$$= \frac{3a \cdot (a-1)}{(a+4)(a+1) \cdot (a-1)} + \frac{2a \cdot (a+4)}{(a+1)(a-1) \cdot (a+4)}$$

$$= \frac{3a(a-1) + 2a(a+4)}{(a+4)(a+1)(a-1)}$$

$$= \frac{3a^2 - 3a + 2a^2 + 8a}{(a+4)(a+1)(a-1)}$$

$$= \frac{5a^2 + 5a}{(a+4)(a+1)(a-1)}$$

$$\frac{5a(a+1)}{(a+4)(a+1)(a-1)} = \frac{5a}{(a+4)(a-1)}$$

27. $\dfrac{6x}{5} \cdot \dfrac{10}{18x^2} = \dfrac{\overset{1}{\cancel{6}} \overset{1}{\cancel{x}} \cdot \overset{2}{\cancel{10}}}{\underset{1}{\cancel{5}} \cdot \underset{3}{\cancel{18}} \underset{x}{\cancel{x^2}}} = \dfrac{2}{3x}$

29.
$$\frac{2x+6}{x^2-9} \cdot \frac{x^2-3x}{4} = \frac{2(x+3)}{(x+3)(x-3)} \cdot \frac{x(x-3)}{4}$$

$$= \frac{\overset{1}{\cancel{2}}(\overset{1}{\cancel{x+3}}) \cdot x(\overset{1}{\cancel{x-3}})}{(\underset{1}{\cancel{x+3}})(\underset{1}{\cancel{x-3}}) \cdot \underset{2}{\cancel{4}}}$$

$$= \frac{x}{2}$$

31. $\dfrac{3p}{5} \div \dfrac{9p^2}{10} = \dfrac{3p}{5} \cdot \dfrac{10}{9p^2} = \dfrac{\overset{1}{\cancel{3}} \overset{1}{\cancel{p}} \cdot \overset{2}{\cancel{10}}}{\underset{1}{\cancel{5}} \cdot \underset{3}{\cancel{9}} \underset{p}{\cancel{p^2}}} = \dfrac{2}{3p}$

33.
$$\frac{x^2+7x+10}{x^2+5x} \div \frac{x^2-4}{2x^2-7x+6}$$

$$= \frac{x^2+7x+10}{x^2+5x} \cdot \frac{2x^2-7x+6}{x^2-4}$$

$$= \frac{(x+5)(x+2)}{x(x+5)} \cdot \frac{(2x-3)(x-2)}{(x+2)(x-2)}$$

$$= \frac{(\overset{1}{\cancel{x+5}})(\overset{1}{\cancel{x+2}}) \cdot (2x-3)(\overset{1}{\cancel{x-2}})}{x(\underset{1}{\cancel{x+5}}) \cdot (\underset{1}{\cancel{x+2}})(\underset{1}{\cancel{x-2}})}$$

$$= \frac{2x-3}{x}$$

35.
$$\frac{a^2b + 2ab^2}{a^2 - 4b^2} \div \frac{4a^2b}{a^2 - ab - 2b^2}$$

$$= \frac{a^2b + 2ab^2}{a^2 - 4b^2} \cdot \frac{a^2 - ab - 2b^2}{4a^2b}$$

$$= \frac{ab(a+2b)}{(a-2b)(a+2b)} \cdot \frac{(a-2b)(a+b)}{4a^2b}$$

$$= \frac{\overset{1}{\cancel{a}}\,\overset{1}{\cancel{b}}\,(\overset{1}{\cancel{a+2b}}) \cdot (\overset{1}{\cancel{a-2b}})(a+b)}{(\underset{1}{\cancel{a-2b}})(\underset{1}{\cancel{a+2b}}) \cdot 4\underset{a}{\cancel{a^2}}\,\underset{1}{\cancel{b}}}$$

$$= \frac{a+b}{4a}$$

37. $\dfrac{\frac{x^2}{12}}{\frac{x^3}{8}} = \dfrac{x^2}{12} \cdot \dfrac{8}{x^3} = \dfrac{2}{3x}$

39. $\dfrac{1 + \frac{x}{y}}{1 - \frac{x}{y}} = \dfrac{\left(1 + \frac{x}{y}\right) \cdot y}{\left(1 - \frac{x}{y}\right) \cdot y} = \dfrac{y + x}{y - x}$

41. $\dfrac{\frac{1}{m} - \frac{1}{n}}{\frac{1}{m} + \frac{1}{n}} = \dfrac{\left(\frac{1}{m} - \frac{1}{n}\right) \cdot mn}{\left(\frac{1}{m} + \frac{1}{n}\right) \cdot mn} = \dfrac{nm}{n + m}$

43. $\dfrac{\frac{2}{a+1} + 1}{1 - \frac{4}{a+1}} = \dfrac{\left(\frac{2}{a+1} + 1\right) \cdot (a+1)}{\left(1 - \frac{4}{a+1}\right) \cdot (a+1)} = \dfrac{2 + a + 1}{a + 1 - 4} = \dfrac{a+3}{a-3}$

45. $\dfrac{x}{5}$; None

47. $\dfrac{2}{(x+1)(x-2)}$; Since $(x + 1)(x - 2) = 0$ implies $x = -1$ or $x = 2$, then -1 and 2 are the excluded values.

49. $\dfrac{x-1}{x^2+3x+2} = \dfrac{x-1}{(x+1)(x+2)}$; Since $(x + 1)(x + 2) = 0$ implies $x = -1$ or $x = -2$, then -1 and -2 are the excluded values.

51.

$$\frac{x-3}{8} = \frac{x-2}{10}$$

$$10(x-3) = 8(x-2)$$
$$10x - 30 = 8x - 16$$
$$10x - 8x = 30 - 16$$
$$2x = 14$$
$$x = 7$$

53. $\dfrac{x}{4} - \dfrac{x}{5} = 2$

The LCD is 20.

$$20 \cdot \frac{x}{4} - 20 \cdot \frac{x}{5} = 20 \cdot 2$$
$$5x - 4x = 40$$
$$x = 40$$

55. $\dfrac{x}{x-2} + 1 = \dfrac{x+4}{x-2}$

The LCD is $x-2$.

$$(x-2) \cdot \frac{x}{x-2} + (x-2) \cdot 1 = (x-2) \cdot \frac{x+4}{x-2}$$
$$x + x - 2 = x + 4$$
$$2x - 2 = x + 4$$
$$x = 6$$

57.

$$\frac{x}{2x-6} - \frac{x-4}{x-3} = \frac{1}{8}$$

$$\frac{x}{2(x-3)} - \frac{x-4}{x-3} = \frac{1}{8}$$

The LCD is $8(x-3)$.

$$8(x-3) \cdot \frac{x}{2(x-3)} - 8(x-3) \cdot \frac{x-4}{x-3} = 8(x-3) \cdot \frac{1}{8}$$
$$4x - 8(x-4) = x - 3$$
$$4x - 8x + 32 = x - 3$$
$$-5x = -35$$
$$x = 7$$

59. $\dfrac{x}{x-5} = \dfrac{3x}{x^2 - 7x + 10} + \dfrac{8}{x-2}$

The LCD is $(x-5)(x-2)$.

$$(x-5)(x-2) \cdot \frac{x}{x-5} = (x-5)(x-2) \cdot \frac{3x}{(x-5)(x-2)} + (x-5)(x-2) \cdot \frac{8}{x-2}$$
$$x(x-2) = 3x + 8(x-5)$$
$$x^2 - 2x = 3x + 8x - 40$$
$$x^2 - 13x + 40 = 0$$
$$(x-8)(x-5) = 0$$
$$x - 8 = 0 \quad \text{or} \quad x - 5 = 0$$
$$x = 8 \quad \text{or} \quad x = 5$$

Since 5 is an excluded value, the solution is 8.

61. $\dfrac{24}{x+2} - 2 = \dfrac{2}{x-3}$

The LCD is $(x+2)(x-3)$.

$$(x+2)(x-3) \cdot \dfrac{24}{x+2} - 2(x+2)(x-3) = (x+2)(x-3) \cdot \dfrac{2}{x-3}$$
$$24(x-3) - 2(x+2)(x-3) = 2(x+2)$$
$$24x - 72 - 2x^2 + 2x + 12 = 2x + 4$$
$$-2x^2 + 26x - 60 = 2x + 4$$
$$0 = 2x^2 - 24x + 64$$
$$0 = 2(x^2 - 12x + 32)$$
$$0 = 2(x-4)(x-8)$$

$$x - 4 = 0 \quad \text{or} \quad x - 8 = 0$$
$$x = 4 \quad \text{or} \qquad x = 8$$

63. Let x be the unknown number.

Then $3x$ is 3 times that number.

$$\dfrac{1}{x} + \dfrac{1}{3x} = \dfrac{1}{3}$$
$$3 + 1 = x$$
$$4 = x$$
$$12 = 3x$$

The numbers are 4 and 12.

65. Let r be his rate on the way there.

Then $r - 8$ is his rate on the way back.

	Distance	Rate	Time
Going	240	r	$\dfrac{240}{r}$
Returning	200	$r-8$	$\dfrac{200}{r-8}$

Going time = Returning time

$$\dfrac{240}{r} = \dfrac{200}{r-8}$$
$$240(r-8) = 200r$$
$$240r - 1920 = 200r$$
$$40r = 1920$$
$$r = 48$$
$$r - 8 = 40$$

His rate going was 48 mi/h and his rate returning was 40 mi/h.

67. Let r be the speed of the plane in still air. Then $r - 20$ is the speed against the wind, and $r + 20$ is the speed with the wind.

	Distance	Rate	Time
Against wind	700	$r - 20$	$\dfrac{700}{r - 20}$
With wind	700	$r + 20$	$\dfrac{700}{r + 20}$

Time against wind + Time with wind = 12 hours

$$\frac{700}{r - 20} + \frac{700}{r + 20} = 12$$

$$700(r + 20) + 700(r - 20) = 12(r - 20)(r + 20)$$

$$700r + 14000 + 700r - 14000 = 12r^2 - 4800$$

$$12r^2 - 1400r - 4800 = 0$$

$$4(3r^2 - 350r - 1200) = 0$$

$$3r^2 - 350r - 1200 = 0$$

$$(3r + 10)(r - 120) = 0$$

$$3r + 10 = 0 \quad \text{or} \quad r - 120 = 0$$

$$r = -\frac{10}{3} \quad \text{or} \quad r = 120$$

Disregard a negative rate.
The plane's speed is 120 mi/h in still air.

69. Let x be the amount of the 40% solution to be added.

	Amount	Strength	Total
10% acid	300	0.10	0.10(300)
40% acid	x	0.40	0.40(x)

$$0.10(300) + 0.40(x) = 0.20(300 + x)$$

$$30 + 0.4x = 60 + 0.2x$$

$$300 + 4x = 600 + 2x$$

$$2x = 300$$

$$x = 150$$

150 mL of the 40% solution should be added.

Chapter 6   An Introduction to Graphing

Exercises 6.1

1. $x + y = 6$
    $4 + 2 = 6$; $(4, 2)$ is a solution.
    $-2 + 4 = 2 \neq 6$; $(-2, 4)$ is not a solution.
    $0 + 6 = 6$; $(0, 6)$ is a solution.
    $-3 + 9 = 6$; $(-3, 9)$ is a solution.

3. $2x - y = 8$
    $2(5) - 2 = 8$; $(5, 2)$ is a solution.
    $2(4) - 0 = 8$; $(4, 0)$ is a solution.
    $2(0) - 8 = -8 \neq 8$; $(0, 8)$ is not a solution.
    $2(6) - 4 = 8$; $(6, 4)$ is a solution.

5. $3x + y = 6$
    $3(2) + 0 = 6$; $(2, 0)$ is a solution.
    $3(2) + 3 = 9 \neq 6$; $(2, 3)$ is not a solution.
    $3(0) + 3 \neq 6$; $(0, 2)$ is not a solution.
    $3(1) + 3 = 6$; $(1, 3)$ is a solution.

7. $2x - 3y = 6$
    $2(0) - 3(2) = -6 \neq 6$; $(0, 2)$ is not a solution.
    $2(3) - 3(0) = 6$; $(3, 0)$ is a solution.
    $2(6) - 3(2) = 6$; $(6, 2)$ is a solution.
    $2(0) - 3(-2) = 6$; $(0, -2)$ is a solution.

9. $3x - 2y = 12$
    $3(4) - 2(0) = 12$; $(4, 0)$ is a solution.
    $3\left(\dfrac{2}{3}\right) - 2(-5) = 12$; $\left(\dfrac{2}{3}, -5\right)$ is a solution.
    $3(0) - 2(6) = -12 \neq 12$; $(0, 6)$ is not a solution.
    $3(5) - 2\left(\dfrac{3}{2}\right) = 12$; $\left(5, \dfrac{3}{2}\right)$ is a solution.

11. $y = 4x$
    $0 = 4(0)$; $(0, 0)$ is a solution.
    $3 \neq 4(1) = 4$; $(1, 3)$ is not a solution.
    $8 = 4(2)$; $(2, 8)$ is a solution.
    $2 \neq 4(8) = 32$; $(8, 2)$ is not a solution.

13. $x = 3$; $(3, 5)$, $(3, 0)$, and $(3, 7)$ are solutions.
    $(0, 3)$ is not a solution.

15. $x + y = 12$

$4 + y = 12$	$x + 5 = 12$	$0 + y = 12$	$x + 0 = 12$
$y = 8$	$x = 7$	$y = 12$	$x = 12$
$(4, 8)$	$(7, 5)$	$(0, 12)$	$(12, 0)$

17. $3x + 6 = 9$

$3(3) + y = 9$	$3x + 9 = 9$	$3x + (-3) = 9$	$3(0) + y = 9$
$y = 0$	$x = 0$	$x = 4$	$y = 9$
$(3, 0)$	$(0, 9)$	$(4, -3)$	$(0, 9)$

19. $5x - y = 15$

$5x - 0 = 15$
$x = 3$
$(3, 0)$

$5(2) - y = 15$
$y = -5$
$(2, -5)$

$5(4) - y = 15$
$y = 5$
$(4, 5)$

$5x - (-5) = 15$
$x = 2$
$(2, -5)$

21. $3x - 2y = 12$

$3x - 2(0) = 12$
$x = 4$
$(4, 0)$

$3x - 2(-6) = 12$
$x = 0$
$(0, -6)$

$3(2) - 2y = 12$
$y = -3$
$(2, -3)$

$3x - 2(3) = 12$
$x = 6$
$(6, 3)$

23. $y = 3x + 9$

$0 = 3x + 9$
$-3 = x$
$(-3, 0)$

$y = 3\left(\dfrac{2}{3}\right) + 9$
$y = 11$
$\left(\dfrac{2}{3}, 11\right)$

$y = 3(0) + 9$
$y = 9$
$(0, 9)$

$y = 3\left(-\dfrac{2}{3}\right) + 9$
$y = 7$
$\left(-\dfrac{2}{3}, 7\right)$

25. $y = 3x - 4$

$y = 3(0) - 4$
$y = -4$
$(0, -4)$

$5 = 3x - 4$
$3 = x$
$(3, 5)$

$0 = 3x - 4$
$\dfrac{4}{3} = x$
$\left(\dfrac{4}{3}, 0\right)$

$y = 3\left(\dfrac{5}{3}\right) - 4$
$y = 1$
$\left(\dfrac{5}{3}, 1\right)$

For exercises 27–37, answers may vary.

27. $x - y = 7$; let $x = 0, 2, 4, 6$

$0 - y = 7$
$y = -7$
$(0, -7)$

$2 - y = 7$
$y = -5$
$(2, -5)$

$4 - y = 7$
$y = -3$
$(4, -3)$

$6 - y = 7$
$y = -1$
$(6, -1)$

29. $2x - y = 6$; let $x = 0, 3, 6, 9$

$2(0) - y = 6$
$y = -6$
$(0, -6)$

$2(3) - y = 6$
$y = 0$
$(3, 0)$

$2(6) - y = 6$
$y = 6$
$(6, 6)$

$2(9) - y = 6$
$y = 12$
$(9, 12)$

31. $x + 4y = 8$; let $x = 8, -4, 0, 4$

$8 + 4y = 8$
$y = 0$
$(8, 0)$

$-4 + 4y = 8$
$y = 3$
$(-4, 3)$

$0 + 4y = 8$
$y = 2$
$(0, 2)$

$4 + 4y = 8$
$y = 1$
$(4, 1)$

33. $2x - 5y = 10$; let $x = -5, 0, 5, 10$

$2(-5) - 5y = 10$
$y = -4$
$(-5, -4)$

$2(0) - 5y = 10$
$y = -2$
$(0, -2)$

$2(5) - 5y = 10$
$y = 2$
$(5, 0)$

$2(0) - 5y = 10$
$y = 2$
$(10, 2)$

35. $y = 2x + 3$; let $x = 0, 1, 2, 3$

$y = 2(0) + 3$
$y = 3$
$(0, 3)$

$y = 2(1) + 3$
$y = 5$
$(1, 5)$

$y = 2(2) + 3$
$y = 7$
$(2, 7)$

$y = 2(3) + 3$
$y = 9$
$(3, 9)$

37. $x = -5$; $(-5, 0), (-5, 1), (-5, 2), (-5, 3)$

39. $x + y + z = 0$

$2 + (-3) + z = 0$
$z = 1$
$(2, -3, 1)$

41. $x + y + z = 0$
$1 + y + 5 = 0$
$\qquad y = -6$
$(1, -6, 5)$

43. $\qquad 2x + y + z = 2$
$2(-2) + y + 1 = 2$
$\qquad\qquad y = 5$
$(-2, 5, 1)$

45. $y = .75x + 8$
$y = .75(2) + 8 = 9.5$
$y = .75(5) + 8 = 11.75$
$y = .75(10) + 8 = 15.5$
$y = .75(15) + 8 = 19.25$
$y = .75(20) + 8 = 23$

The hourly wages for producing
2, 5, 10, 15, and 20 units per hour are
$9.50, $11.75, $15.50, $19.25, and $23.00,
respectively.

47. $A = s^2$
$A = 5^2 = 25$
$A = 10^2 = 100$
$A = 12^2 = 144$
$A = 15^2 = 225$

For squares whose sides are
5 cm, 10 cm, 12 cm, and 15 cm, their areas
are $25\ \text{cm}^2$, $100\ \text{cm}^2$, $144\ \text{cm}^2$, and $225\ \text{cm}^2$,
respectively.

49. $y = 162x + 4365$
$y = 162(1) + 4365 = 4527$
$y = 162(2) + 4365 = 4689$
$y = 162(3) + 4365 = 4851$
$y = 162(4) + 4365 = 5013$
$y = 162(6) + 4365 = 5332$

x	1	2	3	4	6
y	4527	4689	4851	5013	5337

51. Writing exercise

Exercises 6.2

In exercises 1–9, count left or right for the
x coordinate, up or down for the y coordinate.

1. $A(5, 6)$

3. $C(2, 0)$

5. $E(-4, -5)$

7. $S(-5, -3)$

9. $U(-3, 5)$

11–15.

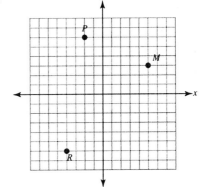

17–21.

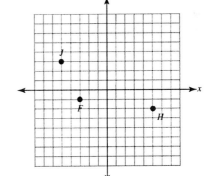

23. The points lie on a line; another point on the line is the point (1, 2).

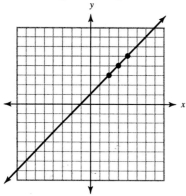

25. The points lie on a line; another point on the line is the point (2, −6).

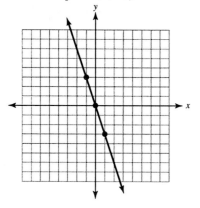

27. **(a)–(c)**
The ordered pairs are A(1500, 350), B(2300, 430), and C(1200, 320).

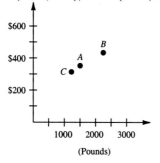

29. The ordered pairs are (1, 4), (2, 14), (3, 26), (4, 33), (5, 42), and (6, 51).

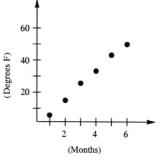

31. The ordered pairs are (1, 4), (2, 7), (3, 6), and (4, 4).

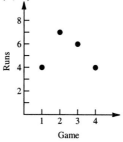

33. Writing exercise

35. **(a)** White Swan: A7; Newport: F2; Wheeler: C2

  (b) A2: Oysterville; F4: Sweet Home; A5: Mineral

Exercises 6.3

1. $x + y = 6$
Two solutions are (0, 6) and (6, 0). The graph is the line through both points.

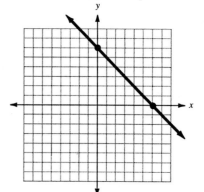

3. $x - y = -3$

Two solutions are $(-3, 0)$ and $(0, 3)$. The graph is the line through both points.

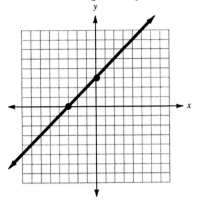

5. $2x + y = 2$

Two solutions are $(0, 2)$ and $(1, 0)$. The graph is the line through both points.

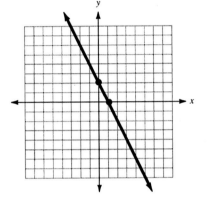

7. $3x + y = 0$

Two solutions are $(0, 0)$ and $(1, -3)$. The graph is the line through both points.

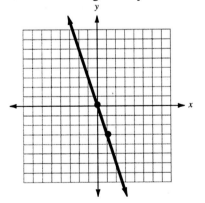

9. $x + 4y = 8$

Two solutions are $(0, 2)$ and $(4, 1)$. The graph is the line through both points.

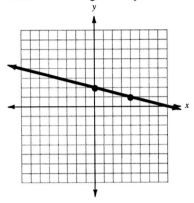

11. $y = 5x$

Two solutions are $(0, 0)$ and $(1, 5)$. The graph is the line through both points.

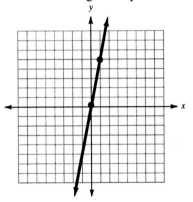

13. $y = 2x - 1$

Two solutions are $(0, -1)$ and $(3, 5)$. The graph is the line through both points.

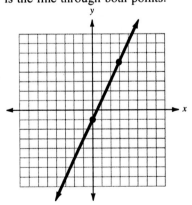

15. $y = -3x + 1$

Two solutions are (0, 1) and (2, −5). The graph is the line through both points.

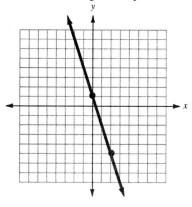

17. $y = \dfrac{1}{3}x$

Two solutions are (0, 0) and (3, 1). The graph is the line through both points.

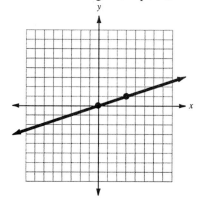

19. $y = \dfrac{2}{3}x - 3$

Two solutions are (0, −3) and (6, 1). The graph is the line through both points.

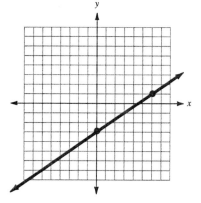

21. $x = 5$

Two solutions are (5, 0) and (5, 6). The graph is the line through both points.

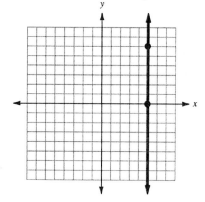

23. $y = 1$

Two solutions are (0, 1) and (4, 1). The graph is the line through both points.

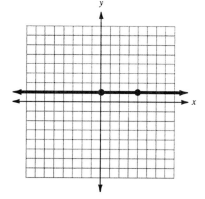

25. $x - 2y = 4$

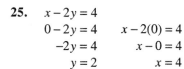

$$\begin{array}{ll} 0 - 2y = 4 & x - 2(0) = 4 \\ -2y = 4 & x - 0 = 4 \\ y = 2 & x = 4 \end{array}$$

(0, 2)   (4, 0)

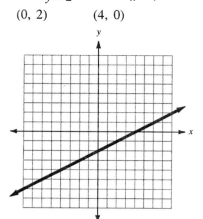

27.
$$5x + 2y = 10$$
$$5(0) + 2y = 10 \qquad 5x + 2(0) = 10$$
$$2y = 10 \qquad\qquad 5x = 10$$
$$y = 5 \qquad\qquad x = 2$$
$$(0,\ 5) \qquad\qquad (2,\ 0)$$

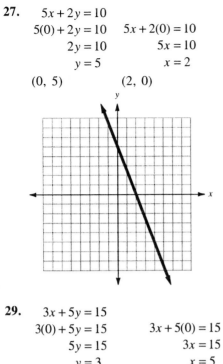

29.
$$3x + 5y = 15$$
$$3(0) + 5y = 15 \qquad 3x + 5(0) = 15$$
$$5y = 15 \qquad\qquad 3x = 15$$
$$y = 3 \qquad\qquad x = 5$$
$$(0,\ 3) \qquad\qquad (5,\ 0)$$

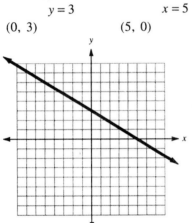

31.
$$x + 3y = 6$$
$$3y = 6 - x$$
$$y = 2 - \frac{1}{3}x$$

Two solutions are $(0, 2)$ and $(6, 0)$. The graph is the line through both points.

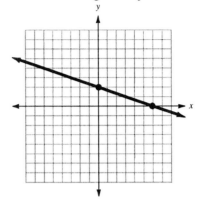

33.
$$3x + 4y = 12$$
$$4y = 12 - 3x$$
$$y = 3 - \frac{3}{4}x$$

Two solutions are $(0, 3)$ and $(4, 0)$. The graph is the line through both points.

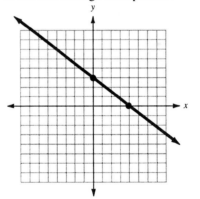

35. $5x - 4y = 20$
$$-4y = 20 - 5x$$
$$y = -5 + \frac{5}{4}x$$

Two solutions are (4, 0) and (0, –5). The graph is the line through both points.

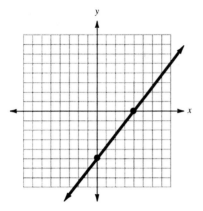

37. $y = 2x$

Two solutions are (0, 0) and (3, 6). The graph is the line through both points.

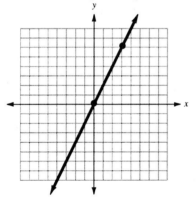

39. $y = x + 3$

Two solutions are (–3, 0) and (0, 3). The graph is the line through both points.

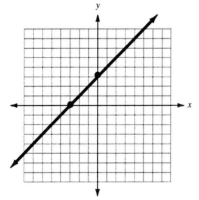

41. $y = 3x - 3$

Two solutions are (0, –3) and (1, 0). The graph is the line through both points.

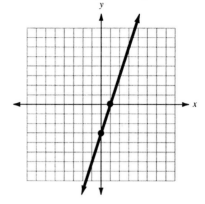

43. $x - 4y = 12$

Two solutions are (0, –3) and (4, –2). The graph is the line through both points.

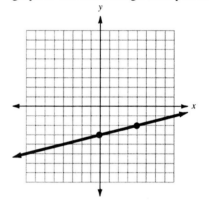
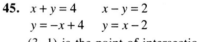

45. $x + y = 4$ $x - y = 2$
      $y = -x + 4$ $y = x - 2$

(3, 1) is the point of intersection.

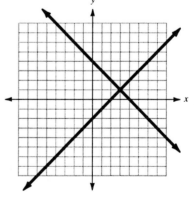

47. Two solutions are (0, 200) and (2000, 400). The graph is the line through both points.

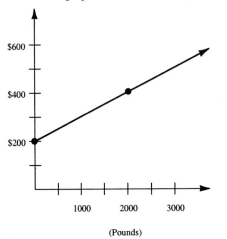

(Pounds)

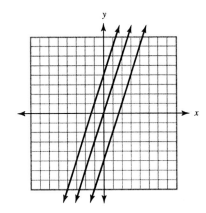

49. (a) Two solutions to the equation $y = 11x - 100$ are (0, −100) and (10, 10). The graph is the line through both points.

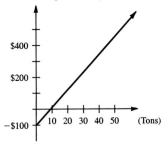

(Tons)

(b) Solve $0 = 11x - 100$.

$$x = \frac{100}{11} \approx 9 \text{ tons}$$

(c) $y = 11(16) - 100 = 76$
The class will make $76.

(d) $y = 17x - 125$

51. $y = 3x \qquad y = 3x + 4 \qquad y = 3x - 5$
$y = 3(0) \qquad y = 3(0) + 4 \qquad y = 3(0) - 5$
$y = 0 \qquad y = 4 \qquad y = -5$

The lines do not intersect. The y intercepts are (0, 0), (0, 4), and (0, −5).

Exercises 6.4

1. (5, 7) and (9, 11)
$$m = \frac{11 - 7}{9 - 5} = \frac{4}{4} = 1$$

3. (−2, −5) and (2, 15)
$$m = \frac{15 - (-5)}{2 - (-2)} = \frac{20}{4} = 5$$

5. (−2, 3) and (3, 7)
$$m = \frac{7 - 3}{3 - (-2)} = \frac{4}{5}$$

7. (−3, 2) and (2, −8)
$$m = \frac{-8 - 2}{2 - (-3)} = \frac{-10}{5} = -2$$

9. (3, 3) and (5, 0)
$$m = \frac{0 - 3}{5 - 3} = -\frac{3}{2}$$

11. (5, −4) and (5, 2)
$$m = \frac{2 - (-4)}{5 - 5} = \frac{6}{0}; \text{ undefined}$$

13. (−4, −2) and (3, 3)
$$m = \frac{3 - (-2)}{3 - (-4)} = \frac{5}{7}$$

15. (−3, −4) and (2, −4)
$$m = \frac{-4 - (-4)}{2 - (-3)} = \frac{0}{5} = 0$$

17. (−1, 7) and (2, 3)
$$m = \frac{3 - 7}{2 - (-1)} = -\frac{4}{3}$$

19. (1, 3) and (2, 5)

$$m = \frac{5-3}{2-1} = \frac{2}{1} = 2$$

21. (3, –1) and (6, –7)

$$m = \frac{-7-(-1)}{6-3} = \frac{-6}{3} = -2$$

23. (4, 5) and (–5, 5)

$$m = \frac{5-5}{-5-4} = \frac{0}{9} = 0$$

25. The line passes through (0, –3) and (1, 1).

$$m = \frac{1-(-3)}{1-0} = \frac{4}{1} = 4$$

27. The line passes through (0, 2) and (1, –3).

$$m = \frac{-3-2}{1-0} = \frac{-5}{1} = -5$$

29. The line passes through (–6, 0) and (0, 2).

$$m = \frac{2-0}{0-(-6)} = \frac{2}{6} = \frac{1}{3}$$

31. $y = -4x$

x	y
0	0
1	–4

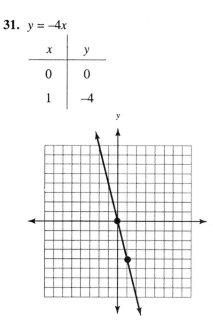

33. $y = \frac{2}{3}x$

x	y
0	0
3	2

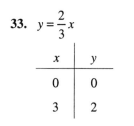

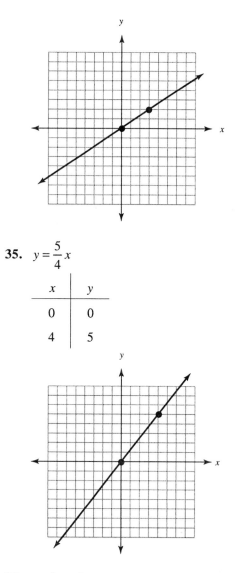

35. $y = \frac{5}{4}x$

x	y
0	0
4	5

37. $y = 2x - 5$

(a)

x	y	
3	1	$y = 2(3) - 5 = 6 - 5 = 1$
4	3	$y = 2(4) - 5 = 8 - 5 = 3$

(b) $m = \frac{3-1}{4-3} = \frac{2}{1} = 2$

(c) The slope equals the coefficient of x.

39. $y = -\dfrac{1}{3}x + 2$

(a)

x	y
3	1	$y = -\dfrac{1}{3}(3) + 2 = -1 + 2 = 1$
6	0	$y = -\dfrac{1}{3}(6) + 2 = -2 + 2 = 0$

(b) $m = \dfrac{0-1}{6-3} = -\dfrac{1}{3}$

(c) The slope equals the coefficient of x.

41. $y = 2x + 3$

(a)

Point	x	y	
A	5	13	$y = 2(5) + 3 = 13$
B	6	15	$y = 2(6) + 3 = 15$
C	7	17	$y = 2(7) + 3 = 17$
D	8	19	$y = 2(8) + 3 = 19$
E	9	21	$y = 2(9) + 3 = 21$

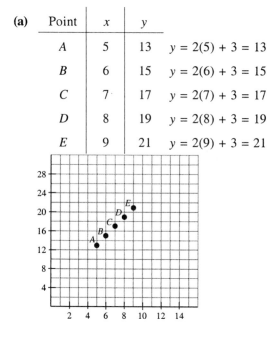

(b) It changes by 2.

43. $y = -4x + 50$

(a)

Point	x	y	
A	5	30	$y = -4(5) + 50 = 30$
B	6	26	$y = -4(6) + 50 = 26$
C	7	22	$y = -4(7) + 50 = 22$
D	8	18	$y = -4(8) + 50 = 18$
E	9	14	$y = -4(9) + 50 = 14$

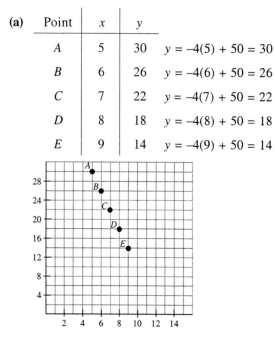

(b) It changes by –4.

(c) Yes

(d) Decreases by 4

45. $(3, 1)$, $m = 2$

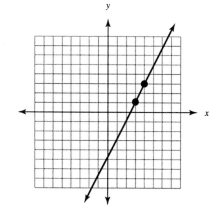

47. $(-2, -1)$, $m = -4$

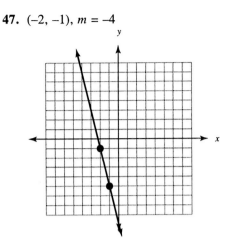

Exercises 6.5

1. $S = 12h$

3. $D = 55h$

5. $y = kx$
   $54 = k \cdot 6$
   $k = 9$

7. $V = kh$
   $189 = k \cdot 9$
   $k = 21$

9. $y = kx$
   $2100 = k \cdot 600$
   $k = 3.5$

11. $k = 2$, $y = 2x$

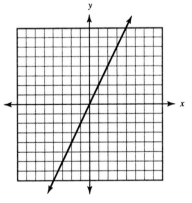

13. $k = 2.5$, $y = 2.5x$

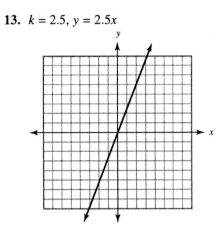

15. $k = 100$, $y = 100x$

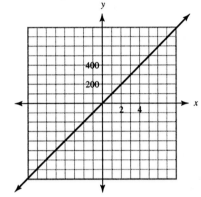

17. $k = 50$, $y = 50x$

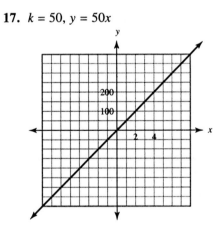

19. $k = 0.20, y = 0.20x$

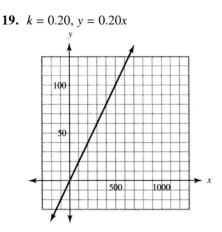

21. $k = 2.50, y = 2.50x$

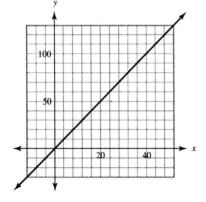

23. Let x be her hours worked and let y be her weekly salary.

$$y = kx$$
$$43.20 = k \cdot 8$$
$$k = 5.40$$

Josephine makes \$5.40 an hour.

$$118.80 = 5.40 \cdot x$$
$$x = 22$$

Josephine worked 22 hours to make \$118.80

Summary Exercises for Chapter 6

1. $7(x) + 2 = 16, x = 2$
$$7(2) + 2 \stackrel{?}{=} 16$$
$$16 = 16$$
Yes, 2 is a solution of $7x + 2 = 16$.

3. $7x - 2 = 2x + 8, x = 2$
$$7(2) - 2 \stackrel{?}{=} 2(2) + 8$$
$$12 = 12$$
Yes, 2 is a solution of $7x - 2 = 2x + 8$.

5. $x + 5 + 3x = 2 + x + 23, x = 6$
$$(6) + 5 + 3(6) \stackrel{?}{=} 2 + (6) + 23$$
$$29 \neq 31$$

No, 6 is not a solution of
$x + 5 + 3x = 2 + x + 23$.

7. $x - y = 6$

(6, 0)	(3, 3)	(3, −3)	(0, −6)
$6 - 0 \stackrel{?}{=} 6$	$3 - 3 \stackrel{?}{=} 6$	$3 - (-3) \stackrel{?}{=} 6$	$0 - (-6) \stackrel{?}{=} 6$
$6 = 6$	$6 = 6$	$6 = 6$	$6 = 6$

The solutions are (6, 0), (3, −3), and (0, −6).

9. $2x + 3y = 6$

(3, 0)	(6, 2)	(−3, 4)	(0, 2)
$2(3) + 3(0) \stackrel{?}{=} 6$	$2(6) + 3(2) \stackrel{?}{=} 6$	$2(-3) + 3(4) \stackrel{?}{=} 6$	$2(0) + 3(2) \stackrel{?}{=} 6$
$6 = 6$	$18 \neq 6$	$6 = 6$	$6 = 6$

The solutions are (3, 0), (3, 4), and (0, 2).

11. $x + y = 8$

$x = 4$	$y = 8$	$x = 8$	$x = 6$
$4 + y = 8$	$x + 8 = 8$	$8 + y = 8$	$6 + y = 8$
$y = 4$	$x = 0$	$y = 0$	$y = 2$
(4, 4)	(0, 8)	(8, 0)	(6, 2)

13.

$x = 3$	$x = 6$	$y = -4$	$x = -3$
$2(3) + 3y = 6$	$2(6) + 3y = 6$	$2x + 3(-4) = 6$	$2(-3) + 3y = 6$
$6 + 3y = 6$	$12 + 3y = 6$	$2x - 12 = 6$	$-6 + 3y = 6$
$3y = 0$	$3y = -6$	$2x = 18$	$3y = 12$
$y = 0$	$y = -2$	$x = 9$	$y = 4$
(3, 0)	(6, −2)	(9, −4)	(−3, 4)

15. $x + y = 10$
$x = 0$:
$0 + y = 10$
$y = 10$
(0, 10) is a solution.
$x = 2$:
$2 + y = 10$
$y = 8$
(2, 8) is a solution.
$x = 4$:
$4 + y = 10$
$y = 6$
(4, 6) is a solution.
$x = 6$:
$6 + y = 10$
$y = 4$
(6, 4) is a solution.

17. $2x - 3y = 6$
$x = 0$:
$2(0) - 3y = 6$
$-3y = 6$
$y = -2$
$(0, -2)$ is a solution.
$x = 3$:
$2(3) - 3y = 6$
$6 - 3y = 6$
$-3y = 0$
$y = 0$
$(3, 0)$ is a solution.
$x = 6$:
$2(6) - 3y = 6$
$12 - 3y = 6$
$-3y = -6$
$y = 2$
$(6, 2)$ is a solution.
$x = 9$:
$2(9) - 3y = 6$
$18 - 3y = 6$
$-3y = -12$
$y = 4$
$(9, 4)$ is a solution.

19. $(4, 6)$

21. $(-1, -5)$

23–25.

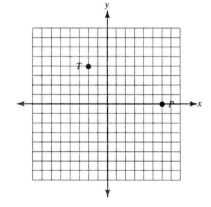

27. $x + y = 5$
Two solutions are $(0, 5)$ and $(5, 0)$. The graph is the line through both points.

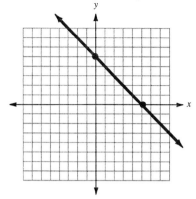

29. $y = 2x$
Two solutions are $(0, 0)$ and $(2, 4)$. The graph is the line through both points.

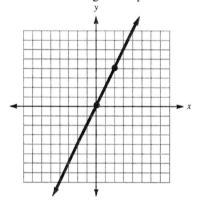

31. $y = \dfrac{3}{2} x$

Two solutions are $(0, 0)$ and $(2, 3)$. The graph is the line through both points.

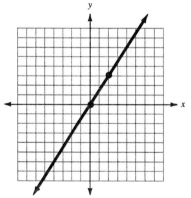

33. $y = 2x - 3$
Two solutions are $(0, -3)$ and $(2, 1)$. The graph is the line through both points.

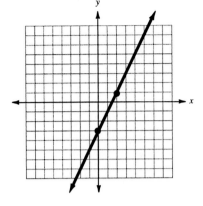

35. $y = \dfrac{2}{3}x + 2$
Two solutions are $(-3, 0)$ and $(0, 2)$. The graph is the line through both points.

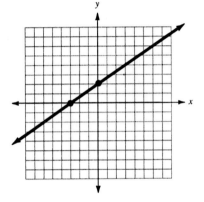

37. $2x + y = 6$
Two solutions are $(0, 6)$ and $(3, 0)$. The graph is the line through both points.

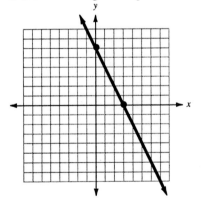

39. $3x - 4y = 12$
Two solutions are $(0, -3)$ and $(4, 0)$. The graph is the line through both points.

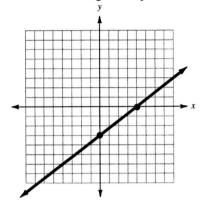

41. $y = -2$
Two solutions are $(0, -2)$ and $(4, -2)$. The graph is the line through both points.

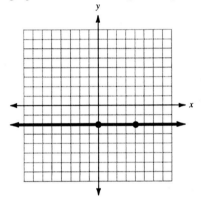

43. $4x + 3y = 12$
Two solutions are $(0, 4)$ and $(3, 0)$. The graph is the line through both points.

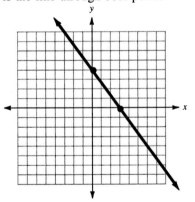

45. $3x + 2y = 6$

$$2y = -3x + 6$$

$$y = -\frac{3}{2}x + 3$$

$x = 0$:

$$y = -\frac{3}{2}(0) + 3$$

$$y = 3$$

So (0, 3) is a solution.

$x = 2$:

$$y = -\frac{3}{2}(2) + 3$$

$$y = -3 + 3$$

$$y = 0$$

So (2, 0) is a solution.

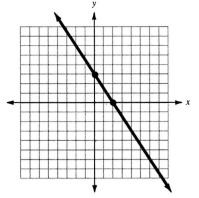

47. (−2, 3) and (1, −6)

$$m = \frac{-6 - 3}{1 - (-2)} = \frac{-9}{3} = -3$$

49. (−5, −2) and (1, 2)

$$m = \frac{2 - (-2)}{1 - (-5)} = \frac{4}{6} = \frac{2}{3}$$

51. (−3, 2) and (−1, −3)

$$m = \frac{-3 - 2}{-1 - (-3)} = \frac{-5}{2}$$

53. (−6, −2) and (−6, 3)

$$m = \frac{3 - (-2)}{-6 - (-6)} = \frac{5}{0} = \text{undefined}$$

55. The line passes through (−2, 1) and (0, −3).

$$m = \frac{-3 - 1}{0 - (-2)} = \frac{-4}{2} = -2$$

57. The line passes through (−6, 0) and (0, −4).

$$m = \frac{-4 - 0}{0 - (-6)} = \frac{-4}{6} = -\frac{2}{3}$$

59. $y = -6x$

x	y
0	0
1	−6

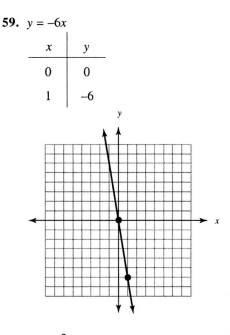

61. $y = -\frac{3}{4}x$

x	y
0	0
4	−3

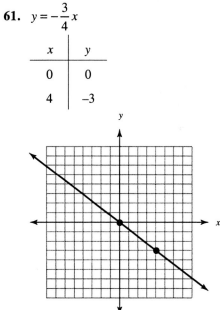

63. $y = kx$

$$5 = k \cdot 3$$

$$k = \frac{5}{3}$$

65. $k = 3.5$, $y = 3.5x$

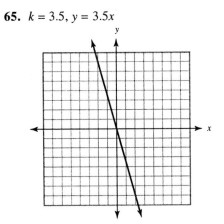

Chapter 7 Graphing and Inequalities

Exercises 7.1

1. $y = 3x + 5$
$m = 3, b = 5$
The slope is 3 and the y intercept is $(0, 5)$.

3. $y = -2x - 5$
$m = -2, b = -5$
The slope is -2 and the y intercept is $(0, -5)$.

5. $y = \dfrac{3}{4}x + 1$

$m = \dfrac{3}{4}, b = 1$

The slope is $\dfrac{3}{4}$ and the y intercept is $(0, 1)$.

7. $y = \dfrac{2}{3}x$

$m = \dfrac{2}{3}, b = 0$

The slope is $\dfrac{2}{3}$ and the y intercept is $(0, 0)$.

9. $4x + 3y = 12$
$3y = -4x + 12$
$y = -\dfrac{4}{3}x + 4$

$m = -\dfrac{4}{3}, b = 4$

The slope is $-\dfrac{4}{3}$ and the y intercept is $(0, 4)$.

11. $y = 9$
$m = 0, b = 9$
The slope is 0 and the y intercept is $(0, 9)$.

13. $3x - 2y = 8$
$-2y = -3x + 8$
$y = \dfrac{3}{2}x - 4$

$m = \dfrac{3}{2}, b = -4$

The slope is $\dfrac{3}{2}$ and the y intercept is $(0, -4)$.

15. $m = 3, b = 5$
$y = mx + b$
$y = 3x + 5$

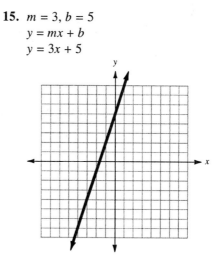

17. $m = -3, b = 4$
$y = mx + b$
$y = -3x + 4$

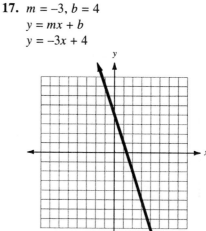

19. $m = \dfrac{1}{2}, b = -2$

$y = mx + b$
$y = \dfrac{1}{2}x - 2$

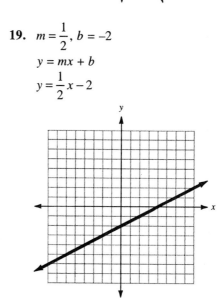

21. $m = -\dfrac{2}{3}$, $b = 0$

$y = mx + b$

$y = -\dfrac{2}{3}x$

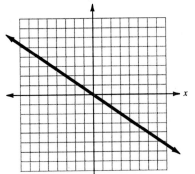

23. $m = \dfrac{3}{4}$, $b = 3$

$y = mx + b$

$y = \dfrac{3}{4}x + 3$

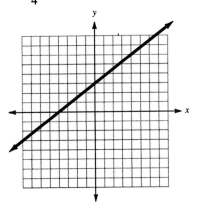

25. slope $= -\dfrac{3}{4}$

y intercept $= (0, 1)$

$y = -\dfrac{3}{4}x + 1$

(g)

27. slope $= -3$

y intercept $= (0, -2)$

$y = -3x - 2$

(e)

29. slope $= -4$

y intercept $= (0, 0)$

$y = -4x$

(h)

31. slope $= -1$

y intercept $= (0, 3)$

$y = -x + 3$

(c)

33. $y = 2x + 1$

Graph the line:

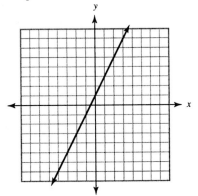

There are no solutions in quadrant IV.

35. $y = -x + 1$

Graph the line:

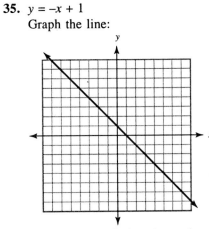

There are no solutions in quadrant III.

37. $y = -2x - 5$

Graph the line:

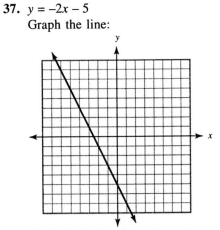

There are no solutions in quadrant I.

39. $y = 3$
Graph the line:

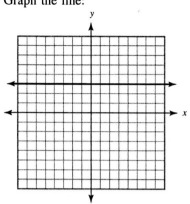

There are no solutions in quadrants III and IV.

41. $y = 0.10x + 200$
slope = 0.10
y intercept = (0, 200)

43. Hourly rate of change $= \dfrac{\text{change in temperature}}{\text{change in hours}}$
$= \dfrac{16°F}{8\ h}$
$= \dfrac{2°F}{h}$

45. slope $= \dfrac{\text{vertical change}}{\text{horizontal change}}$
$= \dfrac{-24,000\ ft}{15\ mi} \cdot \dfrac{1\ mi}{5280\ ft}$
≈ -0.30

47. Writing exercise

49. Group exercise

51.

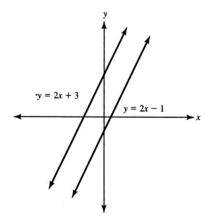

The lines are parallel. They will never intersect.

53.

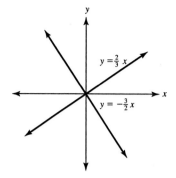

The lines are perpendicular.
$\dfrac{2}{3} \cdot \left(-\dfrac{3}{2}\right) = -1$

55. Find a line perpendicular to $y = \dfrac{3}{5}x.$
$\dfrac{3}{5} \cdot \left(-\dfrac{5}{3}\right) = -1$
$y = -\dfrac{5}{3}x$

Exercises 7.2

1. Because (a) and (d) both have a slope of –4, the lines are parallel.

3. Because (c) and (d) both have a slope of 4, the lines are parallel.

5. Because the product of the slopes for (a) and (c) is $6\left(-\dfrac{1}{6}\right) = -1$, these two lines are perpendicular.

7. Because the product of the slopes for (b) and (d) is $3\left(-\dfrac{1}{3}\right) = -1$, these two lines are perpendicular.

9. $m_1 = \dfrac{3-(-3)}{4-(-2)} = \dfrac{6}{6} = 1$
$m_2 = \dfrac{7-5}{5-3} = \dfrac{2}{2} = 1$
Because the slopes are equal, the lines are parallel.

11. $m_1 = \dfrac{-2-5}{3-8} = \dfrac{-7}{-5} = \dfrac{7}{5}$

$m_2 = \dfrac{-1-4}{4-(-2)} = \dfrac{-5}{6} = -\dfrac{5}{6}$

Because the slopes are neither equal nor negative reciprocals, the lines are neither parallel nor perpendicular.

13. $x - 3y = 6$

$\qquad -3y = -x + 6$

$\qquad\qquad y = \dfrac{1}{3}x - 2$

$3x + y = 3$

$\qquad\quad y = -3x + 3$

Because the slopes are negative reciprocals, the lines are perpendicular.

15. $m = \dfrac{5-3}{4-(-2)} = \dfrac{2}{6} = \dfrac{1}{3}$

17. $m = \dfrac{y-2}{4-(-1)} = \dfrac{y-2}{5}$

Since the line has slope 2,

$\dfrac{y-2}{5} = 2$

$y - 2 = 10$

$\qquad y = 12$

19. Starting with the line in quadrant I, going clockwise around the figure:

$m_1 = \dfrac{1-0}{3-0} = \dfrac{1}{3}$

$m_2 = \dfrac{-1-1}{4-3} = \dfrac{-2}{1} = -2$

$m_3 = \dfrac{-2-(-1)}{1-4} = \dfrac{-1}{-3} = \dfrac{1}{3}$

$m_4 = \dfrac{-2-0}{1-0} = -2$

Since opposite sides have the same slope, they are parallel. But adjacent sides do not have slopes which are negative reciprocals, so they are not perpendicular. Therefore, the figure is a parallelogram.

21. Starting with the line in quadrants I and IV, going clockwise around the figure:

$m_1 = \dfrac{-1-2}{4-3} = \dfrac{-3}{1} = -3$

$m_2 = \dfrac{-3-(-1)}{-2-4} = \dfrac{-2}{-6} = \dfrac{1}{3}$

$m_3 = \dfrac{0-(-3)}{-3-(-2)} = \dfrac{3}{-1} = -3$

$m_4 = \dfrac{2-0}{3-(-3)} = \dfrac{2}{6} = \dfrac{1}{3}$

Since every pair of adjacent sides have slopes which are negative reciprocals, each corner is a right angle. Therefore, the figure is a rectangle.

23. The angle on the negative x axis might be a right angle. The slopes of the sides are:

$m_1 = \dfrac{0-2}{-1-0} = \dfrac{-2}{-1} = 2$

$m_2 = \dfrac{-2-0}{3-(-1)} = \dfrac{-2}{4} = -\dfrac{1}{2}$

The slopes are negative reciprocals, so the sides are perpendicular and the figure is a right angle.

25. Find the slope of the existing line:

$m = -2$

Slope of perpendicular line is $\dfrac{1}{2}$.

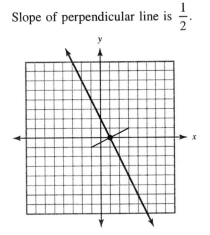

27. Find the slope of the existing line:
$$m = \frac{1}{2}$$
Slope of perpendicular line is –2.

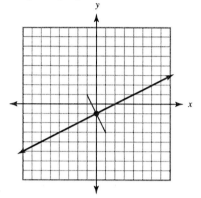

Exercises 7.3

1. $(0, 2)$, $m = 3$
$$y - 2 = 3(x - 0)$$
$$y - 2 = 3x$$
$$y = 3x + 2$$

3. $(0, 2)$, $m = \frac{3}{2}$
$$y - 2 = \frac{3}{2}(x - 0)$$
$$y - 2 = \frac{3}{2}x$$
$$y = \frac{3}{2}x + 2$$

5. $(0, 4)$, $m = 0$
$$y - 4 = 0(x - 0)$$
$$y - 4 = 0$$
$$y = 4$$

7. $(0, -5)$, $m = \frac{5}{4}$
$$y - (-5) = \frac{5}{4}(x - 0)$$
$$y + 5 = \frac{5}{4}x$$
$$y = \frac{5}{4}x - 5$$

9. $(1, 2)$, $m = 3$
$$y - 2 = 3(x - 1)$$
$$y - 2 = 3x - 3$$
$$y = 3x - 1$$

11. $(-2, -3)$, $m = -3$
$$y - (-3) = -3(x - (-2))$$
$$y + 3 = -3(x + 2)$$
$$y + 3 = -3x - 6$$
$$y = -3x - 9$$

13. $(5, -3)$, $m = \frac{2}{5}$
$$y - (-3) = \frac{2}{5}(x - 5)$$
$$y + 3 = \frac{2}{5}x - 2$$
$$y = \frac{2}{5}x - 5$$

15. $(2, -3)$, m is undefined
The line is vertical and of the form $x = a$, where a is the x coordinate of $(2, -3)$.
$$x = 2$$

17. $(2, 3)$ and $(5, 6)$
$$m = \frac{6 - 3}{5 - 2} = \frac{3}{3} = 1$$
$$y - 3 = 1(x - 2)$$
$$y - 3 = x - 2$$
$$y = x + 1$$

19. $(-2, -3)$ and $(2, 0)$
$$m = \frac{0 - (-3)}{2 - (-2)} = \frac{3}{4}$$
$$y - 0 = \frac{3}{4}(x - 2)$$
$$y = \frac{3}{4}x - \frac{3}{2}$$

21. $(-3, 2)$ and $(4, 2)$
$$m = \frac{2 - 2}{4 - (-3)} = \frac{0}{7} = 0$$
$$y - 2 = 0(x - 4)$$
$$y = 2$$

23. $(2, 0)$ and $(0, -3)$
$$m = \frac{-3 - 0}{0 - 2} = \frac{-3}{-2} = \frac{3}{2}$$
$$y - 0 = \frac{3}{2}(x - 2)$$
$$y = \frac{3}{2}x - 3$$

25. $(0, 4)$ and $(-2, -1)$

$$m = \frac{-1-4}{-2-0} = \frac{-5}{-2} = \frac{5}{2}$$

$$y - 4 = \frac{5}{2}(x - 0)$$

$$y - 4 = \frac{5}{2}x$$

$$y = \frac{5}{2}x + 4$$

27. slope 4, y intercept $(0, -2)$

$b = -2$

$y = 4x - 2$

29. x intercept $(4, 0)$ and y intercept $(0, 2)$

$$m = \frac{2-0}{0-4} = \frac{2}{-4} = -\frac{1}{2}$$

$$y - 0 = -\frac{1}{2}(x - 4)$$

$$y = -\frac{1}{2}x + 2$$

31. y intercept $(0, 4)$, slope 0
The line is horizontal and has the form $y = b$,
where b is the y coordinate of $(0, 4)$.
$y = 4$

33. Through $(3, 2)$, slope 5

$y - 2 = 5(x - 3)$

$y - 2 = 5x - 15$

$y = 5x - 13$

35. y intercept $(0, 3)$, parallel to $y = 3x - 5$

$m = 3$

$b = 3$

$y = 3x + 3$

37. y intercept $(0, 4)$, perpendicular to $y = -2x + 1$

$$m = \frac{1}{2}$$

$b = 4$

$$y = \frac{1}{2}x + 4$$

39. y intercept $(0, 3)$, parallel to $y = 2$

$m = 0, b = 3$

$y = 3$

41. Through $(-3, 2)$, parallel to $y = 2x - 3$

$m = 2$

$y - 2 = 2(x - (-3))$

$y - 2 = 2(x + 3)$

$y - 2 = 2x + 6$

$y = 2x + 8$

43. Through $(3, 2)$, parallel to $y = \frac{4}{3}x + 4$

$$m = \frac{4}{3}$$

$$y - 2 = \frac{4}{3}(x - 3)$$

$$y - 2 = \frac{4}{3}x - 4$$

$$y = \frac{4}{3}x - 2$$

45. Through $(5, -2)$, perpendicular to $y = -3x - 2$

$$m = \frac{1}{3}$$

$$y - (-2) = \frac{1}{3}(x - 5)$$

$$y + 2 = \frac{1}{3}x - \frac{5}{3}$$

$$y = \frac{1}{3}x - \frac{11}{3}$$

47. Through $(-2, 1)$, parallel to $x + 2y = 4$

$x + 2y = 4$

$2y = -x + 4$

$$y = -\frac{1}{2}x + 2$$

$$m = -\frac{1}{2}$$

$$y - 1 = -\frac{1}{2}(x - (-2))$$

$$y - 1 = -\frac{1}{2}(x + 2)$$

$$y - 1 = -\frac{1}{2}x - 1$$

$$y = -\frac{1}{2}x$$

49. Writing exercise

51. Find the line that passes through the points (10°C, 50°F) and (40°C, 104°F).

$$m = \frac{104 - 50}{40 - 10} = \frac{54}{30} = \frac{9}{5}$$

$$F - 50 = \frac{9}{5}(C - 10)$$

$$F - 50 = \frac{9}{5}C - 18$$

$$F = \frac{9}{5}C + 32$$

53. Find the line that passes through the points (0, $10,000) and (4, $4000).

$$m = \frac{4000 - 10,000}{4 - 0} = \frac{-6000}{4} = -1500$$

$$V - 10,000 = -1500(t - 0)$$

$$V - 10,000 = -1500t$$

$$V = -1500t + 10,000$$

Exercises 7.4

1. $x + y < 5$
Test (0, 0).
$0 + 0 < 5$
$0 < 5$ A true statement
Shade the half plane containing (0, 0).

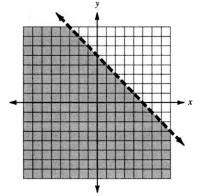

3. $x - 2y \geq 4$
Test (0, 0).
$0 - 2(0) \geq 4$
$0 \geq 4$ A false statement
Shade the half plane that does not contain (0, 0).

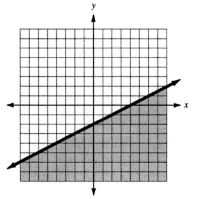

5. $x \leq -3$
Test (0, 0).
$0 \leq -3$ A false statement
Shade the half plane that does not contain (0, 0).

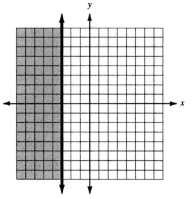

7. $y < 2x - 6$

Test $(0, 0)$

$0 < 2(0) - 6$

$0 < -6$ A false statement

Shade the half plane that does not contain $(0, 0)$.

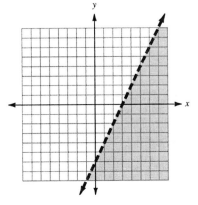

9. $x + y < 3$

$x + y = 3$

$y = -x + 3$

Graph a dashed line with slope -1 and y intercept $(0, 3)$.

Test $(0, 0)$.

$0 + 0 < 3$

$0 < 3$ A true statement

Shade the half plane that contains $(0, 0)$.

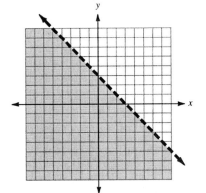

11. $x - y \leq 5$

$x - y = 5$

$-y = -x + 5$

$y = x - 5$

Graph a solid line with slope 1 and y intercept $(0, -5)$.

Test $(0, 0)$.

$0 - 0 \leq 5$

$0 \leq 5$ A true statement

Shade the half plane that contains $(0, 0)$.

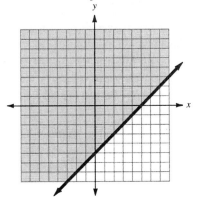

13. $2x + y < 6$

$2x + y = 6$

$y = -2x + 6$

Graph a dashed line with slope -2 and y intercept $(0, 6)$.

Test $(0, 0)$.

$2(0) + 0 < 6$

$0 < 6$ A false statement

Shade the half plane that contains $(0, 0)$.

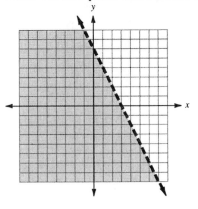

15. $x \leq 3$

Graph a solid vertical line at $x = 3$.
Test $(0, 0)$.
$0 \leq 3$ A true statement
Shade the half plane that contains $(0, 0)$.

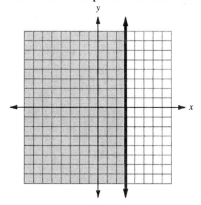

17. $x - 5y < 5$

$x - 5y = 5$

$-5y = -x + 5$

$y = \dfrac{1}{5}x - 1$

Graph a dashed line with slope $\dfrac{1}{5}$ and

y intercept $(0, -1)$.
Test $(0, 0)$.
$0 - 5(0) < 5$

$\qquad 0 < 5$ A true statement
Shade the half plane that contains $(0, 0)$.

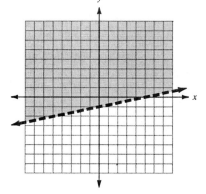

19. $y < -4$

Graph a dashed horizontal line at $y = -4$.
Test $(0, 0)$.
$0 < -4$ A false statement
Shade the half plane that does not contain $(0, 0)$.

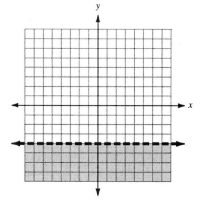

21. $2x - 3y \geq 6$

$2x - 3y = 6$

$-3y = -2x + 6$

$y = \dfrac{2}{3}x - 2$

Graph a solid line with slope $\dfrac{2}{3}$ and

y intercept $(0, -2)$.
Test $(0, 0)$.
$2(0) - 3(0) \geq 6$

$\qquad 0 \geq 6$ A false statement
Shade the half plane that does not contain $(0, 0)$.

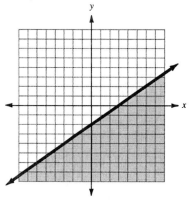

Section 7.4

23. $3x + 2y \geq 0$
$3x + 2y = 0$
$2y = -3x$
$y = -\dfrac{3}{2}x$

Graph a solid line with slope $-\dfrac{3}{2}$ and y

intercept $(0, 0)$.
Test $(1, 1)$.
$3(1) + 2(1) \geq 0$
$3 + 2 \geq 0$
$5 \geq 0$ A true statement
Shade the half plane that contains $(1, 1)$.

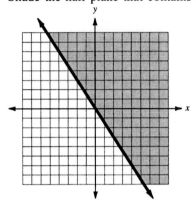

25. $5x + 2y > 10$
$5x + 2y = 10$
$2y = -5x + 10$
$y = -\dfrac{5}{2}x + 5$

Graph a dashed line with slope $-\dfrac{5}{2}$ and

y intercept $(0, 5)$.
Test $(0, 0)$.
$5(0) + 2(0) > 10$
$0 > 10$ A false statement
Shade the half plane that does not contain
$(0, 0)$.

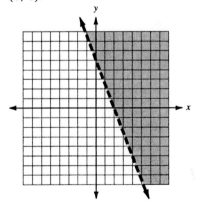

27. $y \leq 2x$
$y = 2x$
Graph a solid line with slope 2 and y intercept
$(0, 0)$.
Test $(1, 1)$.
$1 \leq 2(1)$
$1 \leq 2$ A true statement
Shade the half plane that contains $(1, 1)$.

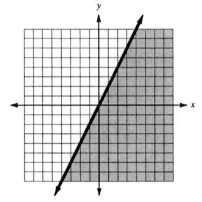

29. $y > 2x - 3$
$y = 2x - 3$
Graph a dashed line with slope 2 and
y intercept $(0, -3)$.
Test $(0, 0)$.
$0 > 2(0) - 3$
$0 > 0 - 3$
$0 > -3$ A true statement
Shade the half plane that contains $(0, 0)$.

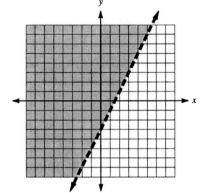

31. $y < -2x - 3$
$y = -2x - 3$
Graph a dashed line with slope -2 and y intercept $(0, -3)$.
Test $(0, 0)$.
$0 < -2(0) - 3$
$0 < 0 - 3$
$0 < -3$ A false statement
Shade the half plane that does not contain $(0, 0)$.

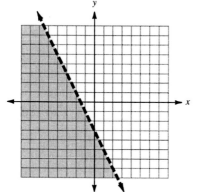

33. $2(x + y) - x > 6$
$2x + 2y - x > 6$
$x + 2y > 6$
$x + 2y = 6$
$2y = -x + 6$
$y = -\frac{1}{2}x + 3$

Graph a dashed line with slope $-\frac{1}{2}$ and y intercept $(0, 3)$.
Test $(0, 0)$.
$0 + 2(0) > 6$
$0 > 6$ A false statement
Shade the half plane that does not contain $(0, 0)$.

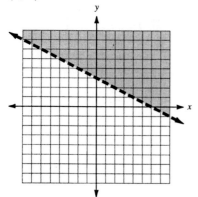

35. $4(x + y) - 3(x + y) \leq 5$
$x + y \leq 5$
$x + y = 5$
$y = -x + 5$
Graph a solid line with slope -1 and y intercept $(0, 5)$.
Test $(0, 0)$.
$0 + 0 \leq 5$
$0 \leq 5$ A true statement
Shade the half plane that contains $(0, 0)$.

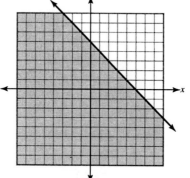

37. Let x be the number of hours worked at the video store, and let y be the number of hours worked at the convenience store.
$9x + 8y \geq 240$

39. Writing exercise

Exercises 7.5

1. $f(x) = x^2 - x - 2$

  (a) $f(0) = 0^2 - 0 - 2 = -2$

  (b) $f(-2) = (-2)^2 - (-2) - 2$
                     $= 4 + 2 - 2$
                     $= 4$

  (c) $f(1) = 1^2 - 1 - 2$
               $= 1 - 1 - 2$
               $= -2$

3. $f(x) = 3x^2 + x - 1$

  (a) $f(-2) = 3(-2)^2 + (-2) - 1$
                 $= 3(4) - 2 - 1$
                 $= 12 - 2 - 1$
                 $= 9$

(b) $f(0) = 3(0)^2 + 0 - 1$
$\quad = -1$

(c) $f(1) = 3(1)^2 + 1 - 1$
$\quad = 3 + 1 - 1$
$\quad = 3$

5. $f(x) = x^3 - 2x^2 + 5x - 2$

    (a) $f(-3) = (-3)^3 - 2(-3)^2 + 5(-3) - 2$
$\quad\quad\quad = -27 - 18 - 15 - 2$
$\quad\quad\quad = -62$

    (b) $f(0) = 0^3 - 2(0)^2 + 5(0) - 2$
$\quad\quad\quad = -2$

    (c) $f(1) = 1^3 - 2(1)^2 + 5(1) - 2$
$\quad\quad\quad = 1 - 2 + 5 - 2$
$\quad\quad\quad = 2$

7. $f(x) = -3x^3 + 2x^2 - 5x + 3$

    (a) $f(-2) = -3(-2)^3 + 2(-2)^2 - 5(-2) + 3$
$\quad\quad\quad = 24 + 8 + 10 + 3$
$\quad\quad\quad = 45$

    (b) $f(0) = -3(0)^3 + 2(0)^2 - 5(0) + 3$
$\quad\quad\quad = 3$

    (c) $f(3) = -3(3)^3 + 2(3)^2 - 5(3) + 3$
$\quad\quad\quad = -81 + 18 - 15 + 3$
$\quad\quad\quad = -75$

9. $f(x) = 2x^3 + 4x^2 + 5x + 2$

    (a) $f(-1) = 2(-1)^3 + 4(-1)^2 + 5(-1) + 2$
$\quad\quad\quad = -2 + 4 - 5 + 2$
$\quad\quad\quad = -1$

    (b) $f(0) = 2(0)^3 + 4(0)^2 + 5(0) + 2$
$\quad\quad\quad = 2$

    (c) $f(1) = 2(1)^3 + 4(1)^2 + 5(1) + 2$
$\quad\quad\quad = 2 + 4 + 5 + 2$
$\quad\quad\quad = 13$

11. $y = -3x + 2$
$\quad f(x) = -3x + 2$

13. $y = 4x - 8$
$\quad f(x) = 4x - 8$

15. $3x + 2y = 6$
$\quad\quad 2y = -3x + 6$
$\quad\quad\quad y = -\dfrac{3}{2}x + 3$
$\quad\quad f(x) = -\dfrac{3}{2}x + 3$

17. $-2x + 6y = 9$
$\quad\quad\quad 6y = 2x + 9$
$\quad\quad\quad y = \dfrac{1}{3}x + \dfrac{3}{2}$
$\quad\quad f(x) = \dfrac{1}{3}x + \dfrac{3}{2}$

19. $-5x - 8y = -9$
$\quad\quad\quad -8y = 5x - 9$
$\quad\quad\quad y = -\dfrac{5}{8}x + \dfrac{9}{8}$
$\quad\quad f(x) = -\dfrac{5}{8}x + \dfrac{9}{8}$

21. $f(x) = 3x + 7$
Find three points:
$f(0) = 3(0) + 7 = 7$
$f(-1) = 3(-1) + 7 = 4$
$f(-2) = 3(-2) + 7 = 1$
Or use the slope and y intercept to graph the line:

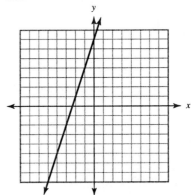

23. $f(x) = -2x + 7$

Find three points:

$f(0) = -2(0) + 7 = 7$
$f(2) = -2(2) + 7 = 3$
$f(4) = -2(4) + 7 = -1$

Or use the slope and y intercept to graph the line:

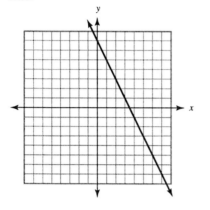

25. $f(x) = -x - 1$

Find three points:

$f(-3) = -(-3) - 1 = 2$
$f(0) = -(0) - 1 = -1$
$f(3) = -3 - 1 = -4$

Or use the slope and y intercept to graph the line:

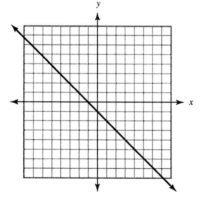

In exercises 27 to 31, $f(x) = 4x - 3$.

27. $f(5) = 4(5) - 3 = 20 - 3 = 17$

29. $f(4) = 4(4) - 3 = 16 - 3 = 13$

31. $f(-4) = 4(-4) - 3 = -16 - 3 = -19$

In exercises 33 to 37, $f(x) = 5x - 1$.

33. $f(a) = 5a - 1$

35. $f(x+1) = 5(x+1) - 1$
$\qquad = 5x + 5 - 1$
$\qquad = 5x + 4$

37. $f(x+h) = 5(x+h) - 1$
$\qquad = 5x + 5h - 1$

In exercises 39 to 41, $g(x) = -3x + 2$.

39. $g(m) = -3m + 2$

41. $g(x+2) = -3(x+2) + 2$
$\qquad = -3x - 6 + 2$
$\qquad = -3x - 4$

In exercises 43 to 45, $f(x) = 2x + 3$.

43. $f(1) = 2(1) + 3 = 5$

45. From exercises 43 and 44, $f(1) = 5$ and $f(3) = 9$.
$(1, f(1)) = (1, 5)$
$(3, f(3)) = (3, 9)$

47. $f(x) = 5x - 2$

(a) $f(4) - f(3) = 5(4) - 2 - (5(3) - 2)$
$\qquad = 20 - 2 - 15 + 2$
$\qquad = 5$

(b) $f(9) - f(8) = 5(9) - 2 - (5(8) - 2)$
$\qquad = 45 - 2 - 40 + 2$
$\qquad = 5$

(c) $f(12) - f(11) = 5(12) - 2 - (5(11) - 2)$
$\qquad = 60 - 2 - 55 + 2$
$\qquad = 5$

(d) The results of (a) through (c) are all 5, which is the same as the slope of the line that is the graph of f.

49. $f(x) = mx + b$

(a) $f(4) - f(3) = m(4) + b - (m(3) + b)$
$\qquad = 4m + b - 3m - b$
$\qquad = m$

(b) $f(9) - f(8) = m(9) + b - (m(8) + b)$

$$= 9m + b - 8m - b$$

$$= m$$

(c) $f(12) - f(11) = m(12) + b - (m(11) + b)$

$$= 12m + b - 11m - b$$

$$= m$$

(d) The results of (a) through (c) are m, which is the slope of the line that is the graph of f.

Summary Exercises for Chapter 7

1. (3, 4) and (5, 8)

$$m = \frac{8-4}{5-3} = \frac{4}{2} = 2$$

3. (–2, 5) and (2, 3)

$$m = \frac{3-5}{2-(-2)} = \frac{-2}{4} = -\frac{1}{2}$$

5. (–2, 6) and (5, 6)

$$m = \frac{6-6}{5-(-2)} = \frac{0}{7} = 0$$

7. (–3, –6) and (5, –2)

$$m = \frac{-2-(-6)}{5-(-3)} = \frac{4}{8} = \frac{1}{2}$$

9. $y = 2x + 5$
$m = 2, b = 5$
slope: 2
y intercept: (0, 5)

11. $y = -\frac{3}{4}x$

$$m = -\frac{3}{4}, \ b = 0$$

slope: $-\frac{3}{4}$

y intercept: (0, 0)

13. $2x + 3y = 6$

$$3y = -2x + 6$$

$$y = -\frac{2}{3}x + 2$$

$$m = -\frac{2}{3}, \ b = 2$$

slope: $-\frac{2}{3}$

y intercept: (0, 2)

15. $y = -3$
$m = 0, \ b = -3$
slope: 0
y intercept: (0, –3)

17. Slope = 2; y intercept: (0, 3)
$y = 2x + 3$

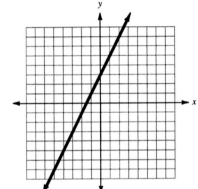

19. Slope $= -\frac{2}{3}$; y intercept: (0, 2)

$$y = -\frac{2}{3}x + 2$$

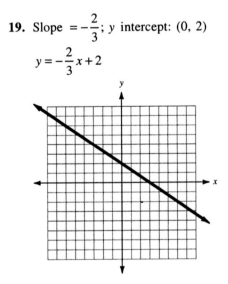

21. $m_1 = \dfrac{-3-1}{2-(-4)} = \dfrac{-4}{6} = -\dfrac{2}{3}$

$m_2 = \dfrac{0-(-3)}{2-0} = \dfrac{3}{2}$

Since the slopes are negative reciprocals, the lines are perpendicular.

23. $4x - 6y = 18$

$\quad -6y = -4x + 18$

$\qquad y = \dfrac{2}{3}x - 3$

$m_1 = \dfrac{2}{3}$

$2x - 3y = 6$

$\quad -3y = -2x + 6$

$\qquad y = \dfrac{2}{3}x - 2$

$m_2 = \dfrac{2}{3}$

Since the slopes are equal, the lines are parallel.

25. $(0, -3)$, $m = 0$

$b = -3$

$y = -3$

27. $(4, 3)$, m undefined

Slope-intercept form is not possible.

$x = 4$

29. $(-2, -3)$, $m = 0$

$b = -3$

$y = -3$

31. $(-3, 2)$, $m = -\dfrac{4}{3}$

$y - 2 = -\dfrac{4}{3}(x - (-3))$

$y - 2 = -\dfrac{4}{3}(x + 3)$

$y - 2 = -\dfrac{4}{3}x - 4$

$\quad y = -\dfrac{4}{3}x - 2$

33. $\left(-\dfrac{5}{2}, -1\right)$, m is undefined

Slope-intercept form is not possible.

$x = -\dfrac{5}{2}$

35. $(0, 4)$ and $(5, 3)$

$m = \dfrac{3-4}{5-0} = \dfrac{-1}{5} = -\dfrac{1}{5}$

$y - 4 = -\dfrac{1}{5}(x - 0)$

$y - 4 = -\dfrac{1}{5}x$

$\quad y = -\dfrac{1}{5}x + 4$

37. $(4, -3)$, slope of $-\dfrac{5}{4}$

$y - (-3) = -\dfrac{5}{4}(x - 4)$

$y + 3 = -\dfrac{5}{4}x + 5$

$\quad y = -\dfrac{5}{4}x + 2$

39. $(3, -2)$, perpendicular to $3x - 5y = 15$

$3x - 5y = 15$

$\quad -5y = -3x + 15$

$\qquad y = \dfrac{3}{5}x - 3$

$m = -\dfrac{5}{3}$

$y - (-2) = -\dfrac{5}{3}(x - 3)$

$y + 2 = -\dfrac{5}{3}x + 5$

$\quad y = -\dfrac{5}{3}x + 3$

41. $(-5, -2)$, parallel to $4x - 3y = 9$

$4x - 3y = 9$

$\quad -3y = -4x + 9$

$\qquad y = \dfrac{4}{3}x - 3$

$m = \dfrac{4}{3}$

$y - (-2) = \dfrac{4}{3}(x - (-5))$

$y + 2 = \dfrac{4}{3}(x + 5)$

$y + 2 = \dfrac{4}{3}x + \dfrac{20}{3}$

$\quad y = \dfrac{4}{3}x + \dfrac{14}{3}$

43. $x - y > 5$

$x - y = 5$

$-y = -x + 5$

$y = x - 5$

Graph a dashed line with slope 1 and y intercept $(0, -5)$.
Test $(0, 0)$.
$0 - 0 > 5$

$0 > 5$ A false statement
Shade the half plane that does not contain $(0, 0)$.

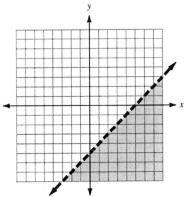

45. $2x - y \geq 6$

$2x - y = 6$

$-y = -2x + 6$

$y = 2x - 6$

Graph a solid line with slope 2 and y intercept $(0, -6)$.
Test $(0, 0)$.
$2(0) - 0 \geq 6$

$0 \geq 6$ A false statement
Shade the half plane that does not contain $(0, 0)$.

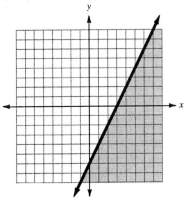

47. $y \leq 2$

Graph a solid horizontal line at $y = 2$.
Test $(0, 0)$.
$0 \leq 2$ A true statement
Shade the half plane that contains $(0, 0)$.

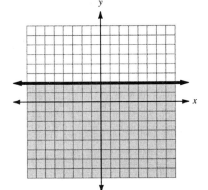

49. $f(x) = -2x^2 + x - 7$

(a) $f(0) = -2(0) + 0 - 7 = -7$

(b) $f(2) = -2(2)^2 + 2 - 7$

$= -8 + 2 - 7$

$= -13$

(c) $f(-2) = -2(-2)^2 + (-2) - 7$

$= -8 - 2 - 7$

$= -17$

51. $f(x) = -x^2 + 7x - 9$

(a) $f(-3) = -(-3)^2 + 7(-3) - 9$

$= -9 - 21 - 9$

$= -39$

(b) $f(0) = -(0)^2 + 7(0) - 9 = -9$

(c) $f(1) = -(1)^2 + 7(1) - 9$

$= -1 + 7 - 9$

$= -3$

53. $f(x) = x^3 + 3x - 5$

(a) $f(2) = 2^3 + 3(2) - 5$

$= 8 + 6 - 5$

$= 9$

(b) $f(0) = (0)^3 + 3(0) - 5 = -5$

(c) $f(1) = 1^3 + 3(1) - 5$
$$= 1 + 3 - 5$$
$$= -1$$

55. $y = -7x - 3$
$f(x) = -7x - 3$

57. $-3x - 2y = 12$
$$-2y = 3x + 12$$
$$y = -\frac{3}{2}x - 6$$
$$f(x) = -\frac{3}{2}x - 6$$

59. $f(x) = 3x - 6$

Find three points, or use the slope of 3 and the y intercept of $(0, -6)$ to graph the line:

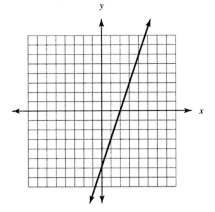

61. $f(x) = -x + 3$

Find three points, or use the slope of -1 and the y intercept of $(0, 3)$ to graph the line:

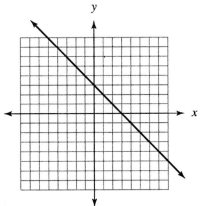

63. $f(x) = -2x + 6$

Find three points, or use the slope of -2 and the y intercept of $(0, 6)$ to graph the line:

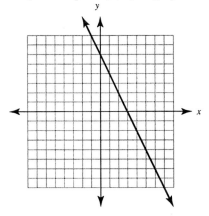

65. $f(x) = -3x + 5$
$f(0) = -3(0) + 5 = 5$
$f(1) = -3(1) + 5$
$$= -3 + 5$$
$$= 2$$

67. $f(x) = -2x + 5$
$f(0) = -2(0) + 5 = 5$
$f(-2) = -2(-2) + 5$
$$= 4 + 5$$
$$= 9$$

69. $f(x) = 7x - 1$
$f(a) = 7a - 1$
$f(3b) = 7(3b) - 1$
$$= 21b - 1$$
$f(x - 1) = 7(x - 1) - 1$
$$= 7x - 7 - 1$$
$$= 7x - 8$$

Chapter 8 Systems of Linear Equations

Exercises 8.1

1. $x + y = 6$
$x - y = 4$

For $x + y = 6$, two solutions are (6, 0) and (0, 6).
For $x - y = 4$, two solutions are (4, 0) and (0, −4).

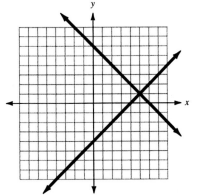

The solution is (5, 1), the intersection point.

3. $-x + y = 3$
$x + y = 5$

For $-x + y = 3$, two solutions are (−3, 0) and (0, 3).
For $x + y = 5$, two solutions are (5, 0) and (0, 5).

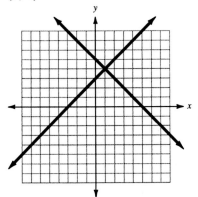

The solution is (1, 4), the intersection point.

5. $x + 2y = 4$
$x - y = 1$

For $x + 2y = 4$, two solutions are (4, 0) and (0, 2).
For $x - y = 1$, two solutions are (1, 0) and (0, −1).

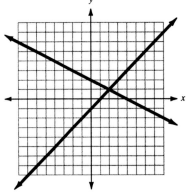

The solution is (2, 1), the intersection point.

7. $2x + y = 8$
$2x - y = 0$

For $2x + y = 8$, two solutions are (4, 0) and (0, 8).
For $2x - y = 0$, two solutions are (0, 0) and (1, 2).

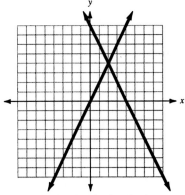

The solution is (2, 4), the intersection point.

9. $x + 3y = 12$
$2x - 3y = 6$

For $x + 3y = 12$, two solutions are (0, 4) and (3, 3).

For $2x - 3y = 6$, two solutions are (3, 0) and (0, –2).

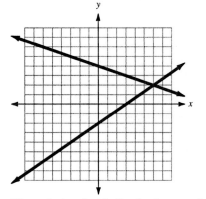

The solution is (6, 2), the intersection point.

11. $3x + 2y = 12$
$y = 3$

For $3x + 2y = 12$, two solution are (4, 0) and (0, 6).

For $y = 3$, two solutions are (0, 3) and (1, 3).

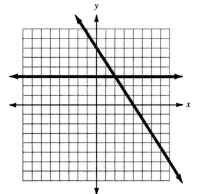

The solution is (2, 3), the intersection point.

13. $x - y = 4$
$2x - 2y = 8$

For $x - y = 4$, two solutions are (4, 0) and (0, – 4).

For $2x - 2y = 8$, two solutions are (4, 0) and (0, – 4).

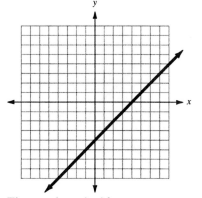

The graphs coincide.
There are infinitely many solutions.
The system is dependent.

15. $x - 4y = -4$
$x + 2y = 8$

For $x - 4y = - 4$, two solutions are (– 4, 0) and (0, 1).

For $x + 2y = 8$, two solutions are (8, 0) and (0, 4).

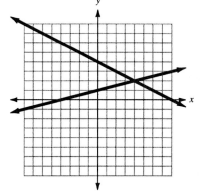

The solution is (4, 2), the intersection point.

17. $3x - 2y = 6$
$2x - y = 5$

For $3x - 2y = 6$, two solutions are $(2, 0)$ and $(0, -3)$.
For $2x - y = 5$, two solutions are $(0, -5)$ and $(2, -1)$.

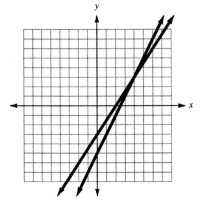

The solution is $(4, 3)$, the intersection point.

19. $3x - y = 3$
$3x - y = 6$

For $3x - y = 3$, two solutions are $(1, 0)$ and $(0, -3)$.
For $3x - y = 6$, two solutions are $(2, 0)$ and $(0, -6)$.

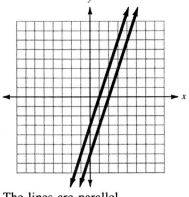

The lines are parallel.
There is no solution.
The system is inconsistent.

21. $2y = 3$
$x - 2y = -3$

For $2y = 3$, two solutions are $\left(0, \dfrac{3}{2}\right)$ and $\left(1, \dfrac{3}{2}\right)$.

For $x - 2y = -3$, two solutions are $(-3, 0)$ and $(-1, 1)$.

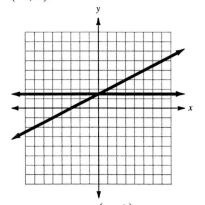

The solution is $\left(0, \dfrac{3}{2}\right)$, the intersection point.

23. $x = 4$
$y = -6$

$x = 4$ is a vertical line with x intercept $(4, 0)$.
$y = -6$ is a horizontal line with y intercept $(0, -6)$.

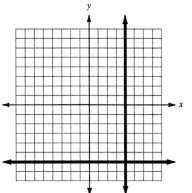

The solution is $(4, -6)$, the intersection point.

25. $mx + 3y = 8$
$-3x + 4y = b$

Substitute $x = 1$ and $y = 2$ in the above equations.

$m(1) + 3(2) = 8$
$m + 6 = 8$
$m = 2$
$-3(1) + 4(2) = b$
$-3 + 8 = b$
$5 = b$

Therefore, $m = 2$ and $b = 5$.

27. Writing exercise

29. Group exercise

Exercises 8.2

1. $x + y = 6$
$\underline{x - y = 4}$
$2x \quad\ = 10$
$x = 5$

Substitute 5 for x in the first equation.
$5 + y = 6$
$y = 1$
So (5, 1) is the solution.

3. $-x + y = 3$
$\underline{x + y = 5}$
$2y = 8$
$y = 4$

Substitute 4 for y in the second equation.
$x + 4 = 5$
$x = 1$

5. $2x - y = 1$
$\underline{-2x + 3y = 5}$
$2y = 6$
$y = 3$

Substitute 3 for y in the first equation.
$2x - 3 = 1$
$2x = 4$
$x = 2$
So (2, 3) is the solution.

7. $x + 3y = 12$
$\underline{2x - 3y = 6}$
$3x \quad\ = 18$
$x = 6$

Substitute 6 for x in the first equation.
$6 + 3y = 12$
$3y = 6$
$y = 2$
So (6, 2) is the solution.

9. $x + 2y = -2$ multiply by -1 $-x - 2y = 2$
$3x + 2y = -12$ $\underline{3x + 2y = -12}$
                                        $2x \quad\ = -10$
                                          $x = -5$

Substitute -5 for x in the first equation.
$-5 + 2y = -2$
$2y = 3$
$y = \dfrac{3}{2}$

So $\left(-5, \dfrac{3}{2}\right)$ is the solution.

11. $4x - 3y = 6$ multiply by -1 $-4x + 3y = -6$
$4x + 5y = 22$ $\underline{4x + 5y = 22}$
                                          $8y = 16$
                                          $y = 2$

Substitute 2 for y in the first equation.
$4x - 3(2) = 6$
$4x = 12$
$x = 3$
So (3, 2) is the solution.

13. $2x + y = 8$ multiply by -1 $-2x - y = -8$
$2x + y = 2$ $\underline{2x + y = 2}$
                                         $0 = -6$

There is no solution.
The system is inconsistent.

15. $3x - 5y = 2$ multiply by -1 $-3x + 5y = -2$
$2x - 5y = -2$ $\underline{2x - 5y = -2}$
                                        $-x \quad\ = -4$
                                          $x = 4$

Substitute 4 for x in the first equation.
$3(4) - 5y = 2$
$-5y = -10$
$y = 2$
So (4, 2) is the solution.

17.
$$x + y = 3 \quad \text{multiply by 2} \quad 2x + 2y = 6$$
$$3x - 2y = 4 \qquad\qquad\qquad\quad \underline{3x - 2y = 4}$$
$$5x \quad\;\; = 10$$
$$x = 2$$

Substitute 2 for x in the first equation.
$$2 + y = 3$$
$$y = 1$$
So (2, 1) is the solution.

19.
$$-5x + 2y = -3 \qquad\qquad\qquad\quad -5x + 2y = -3$$
$$x - 3y = -15 \quad \text{multiply by 5} \quad \underline{5x - 15y = -75}$$
$$-13y = -78$$
$$y = 6$$

Substitute 6 for y in the second equation.
$$x - 3(6) = -15$$
$$x = 3$$
So (3, 6) is the solution.

21.
$$7x + y = 10 \quad \text{multiply by } -3 \quad -21x - 3y = -30$$
$$2x + 3y = -8 \qquad\qquad\qquad\qquad\quad \underline{2x + 3y = -8}$$
$$-19x \quad\;\; = -38$$
$$x = 2$$

Substitute 2 for x in the first equation.
$$7(2) + y = 10$$
$$y = -4$$
So (2, –4) is the solution.

23.
$$5x + 2y = 28 \qquad\qquad\qquad\qquad\quad 5x + 2y = 28$$
$$x - 4y = -23 \quad \text{multiply by } -5 \quad \underline{-5x + 20y = 115}$$
$$22y = 143$$
$$y = \frac{13}{2}$$

Substitute $\frac{13}{2}$ for y in the second equation.
$$x - 4\left(\frac{13}{2}\right) = -23$$
$$x = 3$$
So $\left(3, \frac{13}{2}\right)$ is the solution.

25.
$$3x - 4y = 2 \quad \text{multiply by 2} \quad 6x - 8y = 4$$
$$-6x + 8y = -4 \qquad\qquad\qquad\quad \underline{-6x + 8y = -4}$$
$$0 = 0$$

There are infinitely many solutions.
The system is dependent.

27.
$$5x - 2y = 31 \quad \text{multiply by 3} \quad 15x - 6y = 93$$
$$4x + 3y = 11 \quad \text{multiply by 2} \quad \underline{8x + 6y = 22}$$
$$23x \quad\;\; = 115$$
$$x = 5$$

Substitute 5 for x in the second equation.
$$4(5) + 3y = 11$$
$$3y = -9$$
$$y = -3$$
So (5, –3) is the solution.

29.
$$3x - 2y = 12 \quad \text{multiply by } -3 \quad -9x + 6y = -36$$
$$5x - 3y = 21 \quad \text{multiply by 2} \quad \underline{10x - 6y = 42}$$
$$x \quad\;\; = 6$$

Substitute 6 for x in the first equation.
$$3(6) - 2y = 12$$
$$-2y = -6$$
$$y = 3$$
So (6, 3) is the solution.

31.
$$-2x + 7y = 2 \quad \text{multiply by 3} \quad -6x + 21y = 6$$
$$3x - 5y = -14 \quad \text{multiply by 2} \quad \underline{6x - 10y = -28}$$
$$11y = -22$$
$$y = -2$$

Substitute –2 for y in the first equation.
$$-2x + 7(-2) = 2$$
$$-2x = 16$$
$$x = -8$$
So (–8, –2) is the solution.

33.
$$7x + 4y = 20 \quad \text{multiply by } -3 \quad -21x - 12y = -60$$
$$5x + 6y = 19 \quad \text{multiply by 2} \quad \underline{10x + 12y = 38}$$
$$-11x \quad\;\; = -22$$
$$x = 2$$

Substitute 2 for x in the first equation.
$$7(2) + 4y = 20$$
$$4y = 6$$
$$y = \frac{3}{2}$$
So $\left(2, \frac{3}{2}\right)$ is the solution.

Section 8.2

35. $2x - 7y = 6$ multiply by 2

$-4x + 3y = -12$

$$\begin{array}{r} 4x - 14y = 12 \\ -4x + 3y = -12 \\ \hline -11y = 0 \\ y = 0 \end{array}$$

Substitute 0 for y in the first equation.

$2x - 7(0) = 6$

$2x = 6$

$x = 3$

So $(3, 0)$ is the solution.

37. $5x - y = 20$ multiply by 3

$4x + 3y = 16$

$$\begin{array}{r} 15x - 3y = 60 \\ 4x + 3y = 16 \\ \hline 19x = 76 \\ x = 4 \end{array}$$

Substitute 4 for x in the first equation.

$5(4) - y = 20$

$-y = 0$

$y = 0$

So $(4, 0)$ is the solution.

39. $3x + y = 1$ multiply by -1

$5x + y = 2$

$$\begin{array}{r} -3x - y = -1 \\ 5x + y = 2 \\ \hline 2x = 1 \\ x = \dfrac{1}{2} \end{array}$$

Substitute $\dfrac{1}{2}$ for x in the first equation.

$3\left(\dfrac{1}{2}\right) + y = 1$

$y = -\dfrac{1}{2}$

So $\left(\dfrac{1}{2}, -\dfrac{1}{2}\right)$ is the solution.

41. $3x + 4y = 3$

$6x - 2y = 1$ multiply by 2

$$\begin{array}{r} 3x + 4y = 3 \\ 12x - 4y = 2 \\ \hline 15x = 5 \\ x = \dfrac{1}{3} \end{array}$$

Substitute $\dfrac{1}{3}$ for x in the first equation.

$3\left(\dfrac{1}{3}\right) + 4y = 3$

$4y = 2$

$y = \dfrac{1}{2}$

So $\left(\dfrac{1}{3}, \dfrac{1}{2}\right)$ is the solution.

43. $5x - 2y = \dfrac{9}{5}$ multiply by 2

$3x + 4y = -1$

$$\begin{array}{r} 10x - 4y = \dfrac{18}{5} \\ 3x + 4y = -1 \\ \hline 13x = \dfrac{13}{5} \\ x = \dfrac{1}{5} \end{array}$$

Substitute $\dfrac{1}{5}$ for x in the first equation.

$5\left(\dfrac{1}{5}\right) - 2y = \dfrac{9}{5}$

$-2y = \dfrac{4}{5}$

$y = -\dfrac{2}{5}$

So $\left(\dfrac{1}{5}, -\dfrac{2}{5}\right)$ is the solution.

45. $\dfrac{x}{3} - \dfrac{y}{4} = -\dfrac{1}{2}$ multiply by 12 $4x - 3y = -6$ multiply by 2 $8x - 6y = -12$

$\dfrac{x}{2} - \dfrac{y}{5} = \dfrac{3}{10}$ multiply by 10 $5x - 2y = 3$ multiply by -3 $\dfrac{-15x + 6y = -9}{}$

$$-7x = -21$$
$$x = 3$$

Substitute 3 for x in the equation $4x - 3y = -6$.
$$4(3) - 3y = -6$$
$$-3y = -18$$
$$y = 6$$
So $(3, 6)$ is the solution.

47. $0.4x - 0.2y = 0.6$ multiply by 10 $4x - 2y = 6$ multiply by -3 $-12x + 6y = -18$

$0.5x - 0.6y = 9.5$ multiply by 10 $5x - 6y = 95$ $\dfrac{5x - 6y = 95}{}$

$$-7x = 77$$
$$x = -11$$

Substitute -11 for x in the equation $4x - 2y = 6$.
$$4(-11) - 2y = 6$$
$$-2y = 50$$
$$y = -25$$
So $(-11, -25)$ is the solution.

49. Let x be one number.
Let y be the other number.
$$x + y = 40$$
$$\dfrac{x - y = 8}{}$$
$$2x = 48$$
$$x = 24$$
Substitute 24 for x in the first equation.
$$24 + y = 40$$
$$y = 16$$
The two numbers are 24, 16.

51. Let C be the cost of a chocolate bar.
Let G be the cost of a pack of gum.
$4C + G = 2.75$ multiply by -3 $-12C - 3G = -8.25$
$2C + 3G = 2.25$ $\dfrac{2C + 3G = 2.25}{}$
$$-10C = -6.00$$
$$C = 0.60$$
Substitute 0.60 for C in the first equation.
$$4(0.60) + G = 2.75$$
$$G = 0.35$$
A chocolate bar costs 60¢ and a pack of gum costs 35¢.

53. Let D be the cost of a disk.
Let Z be the cost of a zip disk.

$$8D + 5Z = 27.50$$
$$2D + 4Z = 16.50 \quad \text{multiply by } -4$$

$$8D + 5Z = 27.50$$
$$\underline{-8D - 16Z = -66.00}$$
$$-11Z = -38.50$$
$$Z = 3.50$$

Substitute 3.50 for Z in the first equation.
$$8D + 5(3.50) = 27.50$$
$$8D + 17.50 = 27.50$$
$$8D = 10.00$$
$$D = 1.25$$

A disk costs \$1.25 and a zip disk costs \$3.50.

55. Let x be the length of the shorter piece.
Let y be the length of the longer piece.

$$x + y = 18$$
$$y = 2x$$

$$x + y = 18 \quad \text{multiply by } -1$$
$$-2x + y = 0$$

$$-x - y = -18$$
$$\underline{-2x + y = 0}$$
$$-3x = -18$$
$$x = 6$$

Substitute 6 for x in the second equation.
$y = 2(6) = 12$
The pieces are 6 ft and 12 ft long.

57. Let D be the number of dimes.
Let Q be the number of quarters.

$$D + Q = 22$$
$$0.10D + 0.25Q = 4.00 \quad \text{multiply by } 100$$

$$D + Q = 22 \quad \text{multiply by } -10$$
$$10D + 25Q = 400$$

$$-10D - 10Q = -220$$
$$\underline{10D + 25Q = 400}$$
$$15Q = 180$$
$$Q = 12$$

Substitute 12 for Q in the first equation.
$$D + 12 = 22$$
$$D = 10$$

Richard has 10 dimes and 12 quarters.

59. Let x be the number of \$7 tickets sold.
Let y be the number of \$9 tickets sold.

$$x + y = 400 \quad \text{multiply by } -7$$
$$7x + 9y = 3100$$

$$-7x - 7y = -2800$$
$$\underline{7x + 9y = 3100}$$
$$2y = 300$$
$$y = 150$$

Substitute 150 for y in the first equation.
$$x + 150 = 400$$
$$x = 250$$

There were 250 tickets sold at \$7 and 150 tickets sold at \$9.

61. Let P be the amount of peanuts.
Let C be the amount of cashews.

$$P + C = 20 \qquad P + C = 20 \quad \text{multiply by } -2 \quad -2P - 2C = -40$$
$$2P + 5C = 20(2.75) \qquad 2P + 5C = 55 \qquad\qquad\qquad \underline{2P + 5C = \;\;55}$$
$$3C = \;\;15$$
$$C = \;\;\;5$$

Substitute 5 for C in the first equation.
$$P + 5 = 20$$
$$P = 15$$
15 lb of peanuts and 5 lb of cashews are needed.

63. Let x be the amount of the 30% solution.
Let y be the amount of the 15% solution.

$$x + y = 150 \qquad x + y = 150 \quad \text{multiply by } -0.15 \quad -0.15x - 0.15y = -22.5$$
$$0.30x + 0.15y = 150(0.20) \qquad 0.30x + 0.15y = 30 \qquad\qquad \underline{0.30x + 0.15y = \;\;30}$$
$$0.15x = \;\;7.5$$
$$x = \;\;50$$

Substitute 50 for x in the first equation.
$$50 + y = 150$$
$$y = 100$$
50 mL of the 30% solution and 100 mL of the 15% solution should be used.

65. Let x be the amount invested at 8%.
Let y be the amount invested at 9%.

$$x + y = 12,000 \qquad x + y = 12,000 \quad \text{multiply by } -8 \quad -8x - 8y = -96,000$$
$$0.08x + 0.09y = 1010 \qquad 8x + 9y = 101,000 \qquad\qquad \underline{8x + 9y = 101,000}$$
$$y = \;\;5000$$

Substitute 5000 for y in the first equation.
$$x + 5000 = 12,000$$
$$x = \;\;7000$$
He has \$7000 invested at 8% and \$5000 invested at 9%.

67. Let x be the rate of the plane in still air.
Let y be the rate of the wind.

$$450 = (x + y) \cdot 3 \qquad 3x + 3y = 450 \quad \text{multiply by } 5 \quad 15x + 15y = 2250$$
$$450 = (x - y) \cdot 5 \qquad 5x - 5y = 450 \quad \text{multiply by } 3 \quad \underline{15x - 15y = 1350}$$
$$30x = 3600$$
$$x = \;\;120$$

Substitute 120 for x in the equation $3x + 3y = 450$.
$$3(120) + 3y = 450$$
$$3y = \;\;90$$
$$y = \;\;30$$
The rate of the plane in still air is 120 mi/h and the rate of the wind is 30 mi/h.

69. Let x be one number.
Let y be the other number.
$$y = 3x - 4$$
$$x + y = 36$$
choice d

71. Let W be the width of the rectangle.
Let L be the length of the rectangle.
$$L = 2W + 3$$
$$2L + 2W = 36$$
choice g

73. Let x be the amount of peanuts.
Let y be the amount of cashews.

$$x + \quad y = 140 \qquad x + \quad y = 140$$
$$2x + 5.50y = 140(3) \qquad 2x + 5.5y = 420$$
choice h

75. Let x be the amount of the 20% solution.
Let y be the amount of the 60% solution.

$$x + \quad y = 200 \qquad\qquad x + \quad y = 200$$
$$0.2x + 0.6y = 200(0.45) \qquad 0.2x + 0.6y = 90$$
choice c

77. Writing exercise

Exercises 8.3

1. $x + y = 10 \qquad x + 4x = 10$
$\qquad\quad y = 4x \qquad\quad 5x = 10$
$\qquad\qquad\qquad\qquad\quad x = 2$
Substitute 2 for x in the equation $y = 4x$.
$y = 4(2)$
$y = 8$
So (2, 8) is the solution.

3. $2x - y = 10 \qquad 2(-2y) - y = 10$
$\qquad\quad x = -2y \qquad\qquad -5y = 10$
$\qquad\qquad\qquad\qquad\qquad\quad y = -2$
Substitute -2 for y in the equation $x = -2y$.
$x = -2(-2)$
$x = 4$
So (4, -2) is the solution.

5. $3x + 2y = 12 \qquad 3x + 2(3x) = 12$
$\qquad\qquad y = 3x \qquad\qquad 9x = 12$
$\qquad\qquad\qquad\qquad\qquad\quad x = \dfrac{4}{3}$

Substitute $\dfrac{4}{3}$ for x in the equation $y = 3x$.

$y = 3\left(\dfrac{4}{3}\right)$

$y = 4$

So $\left(\dfrac{4}{3},\ 4\right)$ is the solution.

7. $x + y = 5 \qquad x + (x - 3) = 5$
$\qquad\quad y = x - 3 \qquad\quad 2x - 3 = 5$
$\qquad\qquad\qquad\qquad\qquad\quad 2x = 8$
$\qquad\qquad\qquad\qquad\qquad\quad\ x = 4$
Substitute 4 for x in the equation $y = x - 3$.
$y = 4 - 3$
$y = 1$
So (4, 1) is the solution.

9. $x - y = 4 \qquad (2y - 2) - y = 4$
$\qquad\quad x = 2y - 2 \qquad\quad y - 2 = 4$
$\qquad\qquad\qquad\qquad\qquad\qquad\ y = 6$
Substitute 6 for y in the equation $x = 2y - 2$.
$x = 2(6) - 2$
$x = 10$
So (10, 6) is the solution.

11. $2x + y = 7 \qquad 2x + (x - 8) = 7$
$\qquad\quad y = x - 8 \qquad\quad 3x - 8 = 7$
$\qquad\qquad\qquad\qquad\qquad\quad 3x = 15$
$\qquad\qquad\qquad\qquad\qquad\quad\ x = 5$
Substitute 5 for x in the equation $y = x - 8$.
$y = 5 - 8$
$y = -3$
So (5, -3) is the solution.

13. $\qquad 2x - 5y = 10 \qquad 2(y + 8) - 5y = 10$
$\quad x - y = 8 \text{ or } x = y + 8 \qquad -3y + 16 = 10$
$\qquad\qquad\qquad\qquad\qquad\qquad\qquad\ -3y = -6$
$\qquad\qquad\qquad\qquad\qquad\qquad\qquad\quad y = 2$
Substitute 2 for y in the equation $x = y + 8$.
$x = 2 + 8$
$x = 10$
So (10, 2) is the solution.

15.
$$3x + 4y = 9 \qquad 3x + 4(3x + 1) = 9$$
$$y - 3x = 1 \text{ or } y = 3x + 1 \qquad 15x + 4 = 9$$
$$15x = 5$$
$$x = \frac{1}{3}$$

Substitute $\frac{1}{3}$ for x in the equation $y = 3x + 1$.
$$y = 3\left(\frac{1}{3}\right) + 1$$
$$y = 2$$
So $\left(\frac{1}{3}, \ 2\right)$ is the solution.

17. $3x - 18y = 4 \qquad 3(6y + 2) - 18y = 4$
$$x = 6y + 2 \qquad 6 = 4$$
There is no solution.
The system is inconsistent.

19. $5x - 3y = 6 \qquad 5x - 3(3x - 6) = 6$
$$y = 3x - 6 \qquad -4x + 18 = 6$$
$$-4x = -12$$
$$x = 3$$
Substitute 3 for x in the equation $y = 3x - 6$.
$$y = 3(3) - 6$$
$$y = 3$$
So $(3, 3)$ is the solution.

21. $8x - 5y = 16 \qquad 8x - 5(4x - 5) = 16$
$$y = 4x - 5 \qquad -12x + 25 = 16$$
$$-12x = -9$$
$$x = \frac{3}{4}$$
Substitute $\frac{3}{4}$ for x in the equation $y = 4x - 5$.
$$y = 4\left(\frac{3}{4}\right) - 5$$
$$y = -2$$
So $\left(\frac{3}{4}, -2\right)$ is the solution.

23.
$$x + 3y = 7 \qquad (y + 3) + 3y = 7$$
$$x - y = 3 \text{ or } x = y + 3 \qquad 4y + 3 = 7$$
$$4y = 4$$
$$y = 1$$
Substitute 1 for y in the equation $x = y + 3$.
$$x = 1 + 3$$
$$x = 4$$
So $(4, 1)$ is the solution.

25.
$$6x - 3y = 9 \qquad 6x - 3(2x - 3) = 9$$
$$2x + y = -3 \text{ or } y = 2x - 3 \qquad 9 = 9$$
There are infinitely many solutions.
The system is dependent.

27. $x - 7y = 3 \text{ or } x = 7y + 3$
$$2x - 5y = 15 \qquad 2(7y + 3) - 5y = 15$$
$$9y + 6 = 15$$
$$9y = 9$$
$$y = 1$$
Substitute 1 for y in the equation $x = 7y + 3$.
$$x = 7(1) + 3$$
$$x = 10$$
So $(10, 1)$ is the solution.

29.
$$4x + 3y = -11 \qquad 4x + 3(-5x - 11) = -11$$
$$5x + y = -11 \text{ or } y = -5x - 11 \qquad -11x - 33 = -11$$
$$-11x = 22$$
$$x = -2$$
Substitute -2 for x in the equation $y = -5x - 11$.
$$y = -5(-2) - 11$$
$$y = -1$$
So $(-2, -1)$ is the solution.

31. $2x + 3y = -6 \qquad 2(3y + 6) + 3y = -6$
$$x = 3y + 6 \qquad 9y + 12 = -6$$
$$9y = -18$$
$$y = -2$$
Substitute -2 for y in the equation $x = 3y + 6$.
$$x = 3(-2) + 6$$
$$x = 0$$
So $(0, -2)$ is the solution.

33. $2x - \ y = 1$
$$\underline{-2x + 3y = 5}$$
$$2y = 6$$
$$y = 3$$
Substitute 3 for y in the first equation.
$$2x - 3 = 1$$
$$2x = 4$$
$$x = 2$$
So $(2, 3)$ is the solution.

35. $6x + 2y = 4 \qquad 6x + 2(-3x + 2) = 4$
$$y = -3x + 2 \qquad 4 = 4$$
There are infinitely many solutions.
The system is dependent.

37. $x + 2y = -2$ multiply by -1 $\quad -x - 2y = 2$

 $3x + 2y = -12$ $ \underline{3x + 2y = -12}$

$ 2x = -10$

$ x = -5$

Substitute -5 for x in the first equation.

$-5 + 2y = -2$

$ 2y = 3$

$ y = \dfrac{3}{2}$

So $\left(-5, \dfrac{3}{2} \right)$ is the solution.

39. $2x - 3y = 14$ multiply by -2 $\quad -4x + 6y = -28$

 $4x + 5y = -5$ $\underline{4x + 5y = -5}$

$ 11y = -33$

$ y = -3$

Substitute -3 for y in the first equation.

$2x - 3(-3) = 14$

$ 2x = 5$

$ x = \dfrac{5}{2}$

So $\left(\dfrac{5}{2}, -3 \right)$ is the solution.

41. $4x - 2y = 0$

$ x = \dfrac{3}{2}$

Substitute $\dfrac{3}{2}$ for x in the first equation.

$4 \left(\dfrac{3}{2} \right) - 2y = 0$

$ -2y = -6$

$ y = 3$

So $\left(\dfrac{3}{2}, 3 \right)$ is the solution.

43. $\dfrac{1}{3}x + \dfrac{1}{2}y = 5$ multiply by 6 $\quad 2x + 3y = 30$ multiply by 4 $\quad 8x + 12y = 120$

 $\dfrac{x}{4} - \dfrac{y}{5} = -2$ multiply by 20 $\quad 5x - 4y = -40$ multiply by 3 $\quad \underline{15x - 12y = -120}$

$ 23x = 0$

$ x = 0$

Substitute 0 for x in the equation $2x + 3y = 30$.

$2(0) + 3y = 30$

$ 3y = 30$

$ y = 10$

So $(0, 10)$ is the solution.

45. $0.4x - 0.2y = 0.6$ multiply by 10 $4x - 2y = 6$ multiply by -3 $-12x + 6y = -18$

$2.5x - 0.3y = 4.7$ multiply by 10 $25x - 3y = 47$ multiply by 2

$$50x - 6y = 94$$
$$\overline{}$$
$$38x \qquad = 76$$
$$x = 2$$

Substitute 2 for x in the equation $4x - 2y = 6$.

$4(2) - 2y = 6$

$\quad -2y = -2$

$\qquad y = 1$

So $(2, 1)$ is the solution.

47. Let x be the 1st number.

Let y be the 2nd number.

$x + y = 100 \qquad x + 3x = 100$

$\quad y = 3x \qquad\qquad 4x = 100$

$\qquad\qquad\qquad\qquad x = 25$

Substitute 25 for x in the equation $y = 3x$.

$y = 3(25)$

$y = 75$

The numbers are 25, 75.

49. Let x be the 1st number.

Let y be the 2nd number.

$x + y = 56 \qquad\quad x + (2x - 4) = 56$

$\quad y = 2x - 4 \qquad\qquad 3x = 60$

$\qquad\qquad\qquad\qquad\quad x = 20$

Substitute 20 for x in the equation $y = 2x - 4$.

$y = 2(20) - 4$

$y = 36$

The numbers are 20, 36.

51. Let x be the smaller number.

Let y be the larger number.

$y - x = 22 \qquad (3x + 2) - x = 22$

$\quad y = 3x + 2 \qquad\qquad 2x = 20$

$\qquad\qquad\qquad\qquad\quad x = 10$

Substitute 10 for x in the equation $y = 3x + 2$.

$y = 3(10) + 2$

$y = 32$

The numbers are 10, 32.

53. Let x be the smaller number.

Let y be the larger number.

$y = 5x \qquad\quad 5x = 2x + 9$

$y = 2x + 9 \qquad 3x = 9$

$\qquad\qquad\qquad\quad x = 3$

Substitute 3 for x in the equation $y = 5x$.

$y = 5(3)$

$y = 15$

The numbers are 3, 15.

55. Let x be the cost of the washer.
Let y be the cost of the dryer.

$x + y = 1200$ $\quad$ $(y + 220) + y = 1200$

$\quad x = y + 220$ $\quad\quad\quad$ $2y = 980$

$\quad\quad\quad\quad\quad\quad\quad\quad\quad$ $y = 490$

Substitute 490 for y in the equation $x = y + 220$.

$x = 490 + 220$

$x = 710$

The washer costs \$710 and the dryer costs \$490.

57. Let x be the cost of the desk.
Let y be the cost of the chair.

$x + y = 850$ $\quad$ $(2y - 50) + y = 850$

$\quad x = 2y - 50$ $\quad\quad\quad$ $3y = 900$

$\quad\quad\quad\quad\quad\quad\quad\quad\quad$ $y = 300$

Substitute 300 for y in the equation $x = 2y - 50$.

$x = 2(300) - 50$

$x = 550$

The desk cost \$550 and the chair cost \$300.

59. Let x be the length of one of the two equal legs.
Let y be the length of the base.

$2x + y = 37$ $\quad$ $2(3y - 6) + y = 37$

$\quad x = 3y - 6$ $\quad\quad\quad$ $7y = 49$

$\quad\quad\quad\quad\quad\quad\quad\quad\quad$ $y = 7$

Substitute 7 for y in the equation $x = 3y - 6$.

$x = 3(7) - 6$

$x = 15$

The lengths of the sides are 7 in., 15 in., 15 in.

Exercises 8.4

For exercises 1 to 20, graph the boundary line of each of the inequalities on the same set of axes. Then choose the appropriate half planes. The set of solutions is the intersection of those regions, which is shaded.

1. $x + 2y \le 4$
$\quad$ $x - y \ge 1$

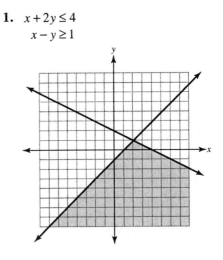

3. $3x + y < 6$
$\quad$ $x + y > 4$

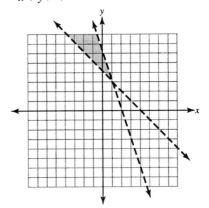

5. $\quad x + 3y \le 12$
$\quad\quad$ $2x - 3y \le 6$

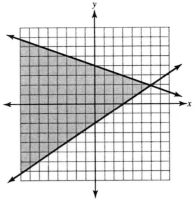

7. $3x + 2y \leq 12$
$\quad x \geq 2$

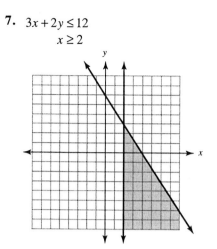

9. $2x + y \leq 8$
$\quad x > 1$
$\quad y > 2$

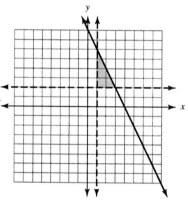

11. $x + 2y \leq 8$
$\quad 2 \leq x \leq 6$
$\quad y \geq 0$

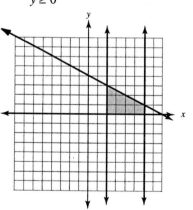

13. $3x + y \leq 6$
$\quad x + y \leq 4$
$\quad x \geq 0$
$\quad y \geq 0$

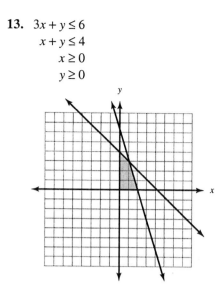

15. $4x + 3y \leq 12$
$\quad x + 4y \leq 8$
$\quad x \geq 0$
$\quad y \geq 0$

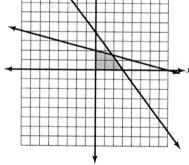

17. $x - 4y \leq -4$
$\quad x + 2y \leq 8$
$\quad x \geq 2$

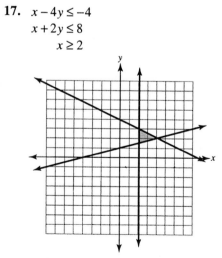

Section 8.4

19. Let x be the number of two-slice toasters.
Let y be the number of four-slice toasters.
$$6x + 10y \le 300$$
$$x + y \le 40$$
$$x \ge 0$$
$$y \ge 0$$

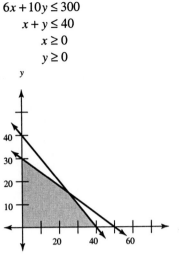

21. Writing exercise

23. $y \le 2x + 3$
$y \le -3x + 5$
$y \ge -x - 1$

Summary Exercises for Chapter 8

1. $x + y = 6$
$x - y = 2$
For $x + y = 6$, two solutions are (6, 0) and (0, 6).
For $x - y = 2$, two solutions are (2, 0) and (0, -2).

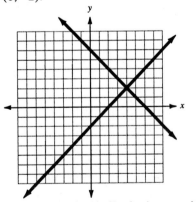

The solution is (4, 2), the intersection point.

3. $x + 2y = 4$
$x + 2y = 6$
For $x + 2y = 4$, two solutions are (4, 0) and

(0, 2).
For $x + 2y = 6$, two solutions are (6, 0) and (0, 3).

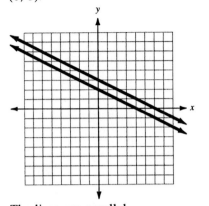

The lines are parallel.
There is no solution.
The system is inconsistent.

5. $2x - 4y = 8$
$x - 2y = 4$
For $2x - 4y = 8$, two solutions are (4, 0) and (0, -2).
For $x - 2y = 4$, two solutions are (4, 0) and (0, -2).

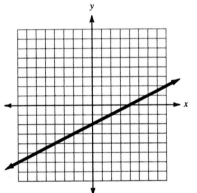

The graphs coincide.
There are infinitely many solutions.
The system is dependent.

7. $x + y = 8$
$\underline{x - y = 2}$
$2x = 10$
$x = 5$

Substitute 5 for x in the first equation.
$5 + y = 8$
$y = 3$
So (5, 3) is the solution.

9. $2x - 3y = 16$

$\dfrac{5x + 3y = 19}{7x \quad\quad = 35}$

$x = 5$

Substitute 5 for x in the first equation.

$2(5) - 3y = 16$

$10 - 3y = 16$

$-3y = 6$

$y = -2$

So $(5, -2)$ is the solution.

11. $3x - 5y = 14$ multiply by -1 $-3x + 5y = -14$

$3x + 2y = 7$ $\dfrac{3x + 2y = \quad 7}{7y = \quad -7}$

$y = -1$

Substitute -1 for y in the first equation.

$3x - 5(-1) = 14$

$3x + 5 = 14$

$3x = 9$

$x = 3$

So $(3, -1)$ is the solution.

13. $4x - 3y = -22$ $\quad\quad\quad\quad\quad\quad 4x - 3y = -22$

$4x + 5y = -6$ multiply by -1 $\dfrac{-4x - 5y = \quad 6}{-8y = -16}$

$y = \quad 2$

Substitute 2 for y in the second equation.

$4x + 5(2) = -6$

$4x + 10 = -6$

$4x = -16$

$x = -4$

So, $(-4, 2)$ is the solution.

15. $4x - 3y = 10$ multiply by -1 $-4x + 3y = -10$

$2x - 3y = 6$ $\dfrac{2x - 3y = \quad 6}{-2x \quad\quad = -4}$

$x = \quad 2$

Substitute 2 for x in the first equation.

$4(2) - 3y = 10$

$8 - 3y = 10$

$-3y = 2$

$y = -\dfrac{2}{3}$

So $\left(2, -\dfrac{2}{3}\right)$ is the solution.

17. $3x + 2y = 3$ multiply by -2 $-6x - 4y = -6$

$6x + 4y = 5$ $\dfrac{6x + 4y = \quad 5}{0 = -1}$

There is no solution.
The system is inconsistent.

19. $5x - 2y = -1$ multiply by -2 $-10x + 4y = \quad 2$

$10x + 3y = 12$ $\dfrac{10x + 3y = 12}{7y = 14}$

$y = \quad 2$

Substitute 2 for y in the first equation.

$5x - 2(2) = -1$

$5x - 4 = -1$

$5x = 3$

$x = \dfrac{3}{5}$

So $\left(\dfrac{3}{5}, \ 2\right)$ is the solution.

21. $2x - 3y = 18$ multiply by -2 $-4x + 6y = -36$

$5x - 6y = 42$ $\dfrac{5x - 6y = \quad 42}{x \quad\quad = \quad 6}$

Substitute 6 for x in the first equation.

$2(6) - 3y = 18$

$12 - 3y = 18$

$-3y = 6$

$y = -2$

So $(6, -2)$ is the solution.

23. $5x - 4y = 12$ multiply by 5 $25x - 20y = \quad 60$

$3x + 5y = 22$ multiply by 4 $\dfrac{12x + 20y = \quad 88}{37x \quad\quad\quad = 148}$

$x = \quad 4$

Substitute 4 for x in the second equation.

$3(4) + 5y = 22$

$12 + 5y = 22$

$5y = 10$

$y = 2$

So $(4, 2)$ is the solution.

25. $4x - 3y = \quad 7$ multiply by 2 $8x - 6y = \quad 14$

$-8x + 6y = -10$ $\dfrac{-8x + 6y = -10}{0 = \quad 4}$

There is no solution.
The system is inconsistent.

27. $3x - 5y = -14$ multiply by -2 $-6x + 10y = 28$

$6x + 3y = -2$ $\underline{6x + 3y = -2}$

                                               $13y = 26$

                                                $y = 2$

Substitute 2 for y in the second equation.

$6x + 3(2) = -2$

$6x + 6 = -2$

$6x = -8$

$x = -\dfrac{4}{3}$

So $\left(-\dfrac{4}{3}, \ 2\right)$ is the solution.

29. $x - y = 10$ $-4y - y = 10$

$ x = -4y$ $-5y = 10$

                           $y = -2$

Substitute -2 for y in the equation $x = -4y$.

$x = -4(-2)$

$x = 8$

So $(8, -2)$ is the solution.

31. $2x + 3y = 2$ $2x + 3(x - 6) = 2$

$ y = x - 6$ $2x + 3x - 18 = 2$

                              $5x = 20$

                                 $x = 4$

Substitute 4 for x in the equation $y = x - 6$.

$y = 4 - 6$

$y = -2$

So $(4, -2)$ is the solution.

33. $x + 5y = 20$ $y + 2 + 5y = 20$

$ x = y + 2$ $6y = 18$

                               $y = 3$

Substitute 3 for y in the equation $x = y + 2$.

$x = 3 + 2$

$x = 5$

So $(5, 3)$ is the solution.

35. $2x + 6y = 10$ $2(6 - 3y) + 6y = 10$

$ x = 6 - 3y$ $12 - 6y + 6y = 10$

                                  $12 = 10$

There is no solution.

The system is inconsistent.

37. $x - 3 = 17$ or $x = 3 + 17$

$2x + y = 6$ $2(3y + 17) + y = 6$

                          $6y + 34 + y = 6$

                               $7y = -28$

                                 $y = -4$

Substitute -4 for y in the equation $x = 3y + 17$.

$x = 3(-4) + 17$

$x = -12 + 17$

$x = 5$

So $(5, -4)$ is the solution.

39. $4x - 5y = -2$

$ x = -3$

Substitute -3 for x in the first equation.

$4(-3) - 5y = -2$

$-12 - 5y = -2$

$-5y = 10$

$y = -2$

So $(-3, -2)$ is the solution.

41. $5x - 2 = -15$ $5x - 2(2x + 6) = -15$

$ y = 2x + 6$ $5x - 5x - 12 = -15$

                                  $x = -3$

Substitute -3 for x in the equation $y = 2x + 6$.

$y = 2(-3) + 6$

$y = -6 + 6$

$y = 0$

So $(-3, 0)$ is the solution.

43. $x - 4y = 0$ $x - 4y = 0$

$4x + y = 34$ multiply by 4 $\underline{16x + 4y = 136}$

                                   $17x = 136$

                                      $x = 8$

Substitute 8 for x in the second equation.

$4(8) + y = 34$

$32 + y = 34$

$y = 2$

So $(8, 2)$ is the solution.

45. $3x - 3y = 30$ $\quad\quad 3(-2y - 8) - 3y = 30$

$\quad\quad x = -2y - 8$ $\quad\quad -6y - 24 - 3y = 30$

$\quad\quad\quad\quad\quad\quad\quad\quad\quad -9y = 54$

$\quad\quad\quad\quad\quad\quad\quad\quad\quad\quad y = -6$

Substitute –6 for y in the equation $x = -2y - 8$.

$x = -2(-6) - 8$

$x = 12 - 8$

$x = 4$

So $(4, -6)$ is the solution.

47. $\quad x - 6y = -8$ $\quad\quad\quad\quad\quad x - 6y = -8$

$\quad 2x + 3y = 4$ multiply by 2 $\quad \underline{4x + 6y = 8}$

$\quad\quad\quad\quad\quad\quad\quad\quad\quad\quad 5x \quad\quad = 0$

$\quad\quad\quad\quad\quad\quad\quad\quad\quad\quad\quad x = 0$

Substitute 0 for x in the second equation.

$2(0) + 3y = 4$

$\quad\quad 3y = 4$

$\quad\quad\quad y = \dfrac{4}{3}$

So $\left(0, \dfrac{4}{3}\right)$ is the solution.

49. $9x + y = 9$ or $y = 9 - 9x$

$\quad x + 3y = 14$ $\quad\quad\quad\quad x + 3(9 - 9x) = 14$

$\quad\quad\quad\quad\quad\quad\quad\quad\quad\quad x + 27 - 27x = 14$

$\quad\quad\quad\quad\quad\quad\quad\quad\quad\quad\quad\quad -26x = -13$

$\quad\quad\quad\quad\quad\quad\quad\quad\quad\quad\quad\quad\quad x = \dfrac{1}{2}$

Substitute $\dfrac{1}{2}$ for x in the equation $y = 9 - 9x$.

$y = 9 - 9\left(\dfrac{1}{2}\right)$

$y = \dfrac{18}{2} - \dfrac{9}{2}$

$y = \dfrac{9}{2}$

So $\left(\dfrac{1}{2}, \dfrac{9}{2}\right)$ is the solution.

51. $3x - 2y = 8$ multiply by -2 $\quad -6x + 4y = -16$

$\quad 2x - 3y = 7$ multiply by 3 $\quad\quad \underline{6x - 9y = 21}$

$\quad\quad\quad\quad\quad\quad\quad\quad\quad\quad\quad -5y = 5$

$\quad\quad\quad\quad\quad\quad\quad\quad\quad\quad\quad\quad y = -1$

Substitute –1 for y in the first equation.

$3x - 2(-1) = 8$

$\quad\quad 3x + 2 = 8$

$\quad\quad\quad 3x = 6$

$\quad\quad\quad\quad x = 2$

So $(2, -1)$ is the solution.

53. Let x be the larger number.
Let y be the smaller number.

$$x + y = 17 \qquad 3xy + 1 + y = 17$$
$$x = 3y + 1 \qquad 4y = 16$$
$$y = 4$$

Substitute 4 for y in the equation $x = 3y + 1$.

$x = 3(4) + 1$
$x = 12 + 1$
$x = 13$
The numbers are 4 and 13.

55. Let x be the cost of a tablet.
Let y be the cost of a pencil.

$$5x + 3y = 8.25 \text{ multiply by } -2 \qquad -10x - 6y = -16.50$$
$$2x + 2y = 3.50 \text{ multiply by } 5 \qquad \underline{10x + 10y = 17.50}$$
$$4y = 1.00$$
$$y = 0.25$$

Substitute 0.25 for y in the first equation.

$5x + 3(0.25) = 8.25$
$5x = 7.50$
$x = 1.50$
Each tablet costs \$1.50 and each pencil costs \$0.25.

57. Let x be the cost of the amplifier.
Let y be the cost of the pair of speakers.

$$x + y = 925 \qquad y + 75 + y = 925$$
$$x = y + 75 \qquad 2y = 850$$
$$y = 425$$

Substitute 425 for y in the equation $x = y + 75$.

$x = 425 + 75$
$x = 500$
The speakers cost \$425 and the amplifier costs \$500.

59. Let x be the length.
Let y be the width.

$$x = y + 4$$
$$2x + 2y = 64 \qquad 2(y + 4) + 2y = 64$$
$$2y + 8 + 2y = 64$$
$$4y = 56$$
$$y = 14$$

Substitute 14 for y in the equation $x = y + 4$.

$x = 14 + 4$
$x = 18$
The length is 18 cm and the width is 14 cm.

61. Let x be the number of nickels.

Let y be the number of quarters.

$$\begin{array}{lll} x + & y = 30 & \text{multiply by } -5 \end{array} \quad \begin{array}{l} -5x - 5y = -150 \end{array}$$

$$\begin{array}{ll} 0.05x + 0.25y = 5.50 & \text{multiply by } 100 \end{array} \quad \underline{\begin{array}{l} 5x + 25y = 550 \end{array}}$$

$$\begin{array}{rr} 20y = & 400 \\ y = & 20 \end{array}$$

Substitute 20 for y in the first equation.

$x + 20 = 30$

$\quad x = 10$

He has 10 nickels and 20 quarters.

63. Let x be the amount of 20% acid solution.

Let y be the amount of 50% acid solution.

$$\begin{array}{ll} x + \quad y = 600 \end{array} \qquad \begin{array}{ll} x + \quad y = 600 & \text{multiply by } -2 \end{array} \quad \begin{array}{l} -2x - 2y = -1200 \end{array}$$

$$\begin{array}{ll} 0.2x + 0.5y = 0.4(600) \end{array} \qquad \begin{array}{ll} 0.2x + 0.5y = 240 & \text{multiply by } 10 \end{array} \quad \underline{\begin{array}{l} 2x + 5y = 2400 \end{array}}$$

$$\begin{array}{rr} 3y = & 1200 \\ y = & 400 \end{array}$$

Substitute 400 for y in the equation $x + y = 600$.

$x + 400 = 600$

$\quad x = 200$

200 mL of 20% acid solution should be mixed with 400 mL of 50% acid solution.

65. Let x be the amount invested at 11%.

Let y be the amount invested at 7%.

$$\begin{array}{lll} x + & y = 18,000 & \text{multiply by } -11 \end{array} \quad \begin{array}{l} -11x - 11y = -198,000 \end{array}$$

$$\begin{array}{ll} 0.11x + 0.07y = 1660 & \text{multiply by } 100 \end{array} \quad \underline{\begin{array}{l} 11x + 7y = 166,000 \end{array}}$$

$$\begin{array}{rr} -4y = & -32,000 \\ y = & 8000 \end{array}$$

Substitute 8000 for y in the first equation.

$x + 8000 = 18,000$

$\quad x = 10,000$

She has $10,000 invested at 11% and $8000 at 7%.

67. Let x be the speed of the plane in still air.

Let y be the speed of the wind.

$$\begin{array}{ll} 2200 = 4(x + y) & \quad 2200 = 4x + 4y \end{array}$$

$$\begin{array}{ll} 1800 = 4(x - y) & \quad \underline{1800 = 4x - 4y} \end{array}$$

$$\begin{array}{l} 4000 = 8x \\ 500 = x \end{array}$$

Substitute 500 for x in the first equation.

$2200 = 4(500 + y)$

$2200 = 2000 + 4y$

$200 = 4y$

$50 = y$

The plane's speed is 500 mi/h and the wind's speed is 50 mi/h.

For exercises 69 to 75, graph the boundary line of each of the inequalities on the same set of axes. Then choose the appropriate half planes. The set of solutions is the intersection of those regions, which is shaded.

69. $x - 2y \le -2$
$x + 2y \le 6$

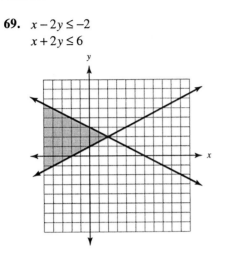

71. $2x + y \le 8$
$x \ge 1$
$y \ge 0$

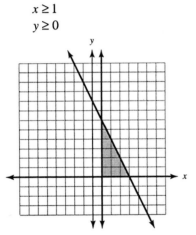

73. $4x + y \le 8$
$x \ge 0$
$y \ge 2$

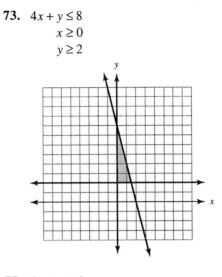

75. $3x + y \le 6$
$x + y \le 4$
$x \ge 0$
$y \ge 0$

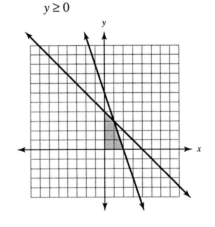

Chapter 9   Exponents and Radicals

Exercises 9.1

1. $\sqrt{16} = \sqrt{4^2} = 4$

3. $\sqrt{400} = \sqrt{20^2} = 20$

5. $-\sqrt{100} = -\sqrt{10^2} = -10$

7. $\sqrt{-81}$ is not a real number, since $-81 < 0$.

9. $\sqrt{\dfrac{16}{9}} = \dfrac{\sqrt{16}}{\sqrt{9}} = \dfrac{4}{3}$

11. Not a real number, since $-\dfrac{4}{5} < 0$

13. $\sqrt[3]{27} = \sqrt[3]{3^3} = 3$

15. $\sqrt[3]{-27} = \sqrt[3]{(-3)^3} = -3$

17. $\sqrt[4]{-81}$ is not a real number, since $-81 < 0$.

19. $-\sqrt[3]{27} = -\sqrt[3]{3^3} = -3$

21. $\sqrt[4]{625} = \sqrt[4]{5^4} = 5$

23. $\sqrt[3]{\dfrac{1}{27}} = \sqrt[3]{\left(\dfrac{1}{3}\right)^3} = \dfrac{1}{3}$

25. $\sqrt{19}$ is irrational because 19 is not a perfect square.

27. $\sqrt{100} = 10$ is rational.

29. $\sqrt[3]{9}$ is irrational because 9 is not a perfect cube.

31. $\sqrt[4]{16} = 2$ is rational.

33. $\sqrt{\dfrac{4}{7}}$ is irrational because $\dfrac{4}{7}$ is not the square of a rational number.

35. $\sqrt[3]{-27} = -3$ is rational.

37. $\sqrt{11} \approx 3.32$

39. $\sqrt{7} \approx 2.65$

41. $\sqrt{46} \approx 6.78$

43. $\sqrt{\dfrac{2}{5}} \approx 0.63$

45. $\sqrt{\dfrac{8}{9}} \approx 0.94$

47. $-\sqrt{18} \approx -4.24$

49. $-\sqrt{27} \approx -5.20$

51. $-\sqrt{16} = -\sqrt{4^2} = -4$

53. $\sqrt[3]{-125} = \sqrt[3]{(-5)^3} = -5 = -\sqrt[3]{5^3} = -\sqrt[3]{125}$

55. $\sqrt[4]{10,000} = \sqrt[4]{10^4} = 10 = \sqrt[3]{10^3} = \sqrt[3]{1000}$

57. $\sqrt{16x^{16}} = \sqrt{(4x^8)^2} = 4x^8 \neq 4x^4$; False

59. True, because $16x^{-4}y^{-4} = \dfrac{16}{x^4y^4} > 0$.

61. $\dfrac{\sqrt{x^2-25}}{x-5} = \dfrac{\sqrt{(x-5)(x+5)}}{x-5} \neq \sqrt{x+5}$; False

63. $A = s^2$
$32 = s^2$
$\sqrt{32} = s$ so $s \approx 5.66$
The length of a side is 5.66 ft.

65. $A = \pi r^2$
$147 = \pi r^2$
$\dfrac{147}{\pi} = r^2$
$\sqrt{\dfrac{147}{\pi}} = r$ so $r \approx 6.84$
The radius is 6.84 ft.

67. $t = \dfrac{1}{4}\sqrt{s} = \dfrac{1}{4}\sqrt{800} \approx 7.07$
It would take 7.07 seconds.

69. $A = s^2$

$10 = s^2$, so

$s = \sqrt{10} \approx 3.16$ ft.

71. $A = s^2$

$17 = s^2$, so

$s = \sqrt{17} \approx 4.12$ ft.

73. Conjugate: $4 + \sqrt{7}$

Opposite: $-\left(4 - \sqrt{7}\right) = -4 + \sqrt{7}$

With the conjugate, only the sign on the radical changes. With the opposite, the sign on the entire expression changes.

75. $\left(\sqrt{3} - 2\right)\left(\sqrt{3} + 2\right)$

$= 3 + 2\sqrt{3} - 2\sqrt{3} - 4$

$= -1$

The answer is $\sqrt{3} + 2$.

77. Challenge exercise

79. (a) $T = 2\pi\sqrt{\dfrac{L}{g}} = 2\pi\sqrt{\dfrac{30}{980}} \approx 1.1$ s

(b) $T = 2\pi\sqrt{\dfrac{L}{g}} = 2\pi\sqrt{\dfrac{50}{980}} \approx 1.4$ s

(c) $T = 2\pi\sqrt{\dfrac{L}{g}} = 2\pi\sqrt{\dfrac{70}{980}} \approx 1.7$ s

(d) $T = 2\pi\sqrt{\dfrac{L}{g}} = 2\pi\sqrt{\dfrac{90}{980}} \approx 1.9$ s

(e) $T = 2\pi\sqrt{\dfrac{L}{g}} = 2\pi\sqrt{\dfrac{110}{980}} \approx 2.1$ s

Exercises 9.2

1. $\sqrt{18} = \sqrt{9 \cdot 2} = \sqrt{9} \cdot \sqrt{2} = 3\sqrt{2}$

3. $\sqrt{28} = \sqrt{4 \cdot 7} = \sqrt{4} \cdot \sqrt{7} = 2\sqrt{7}$

5. $\sqrt{45} = \sqrt{9 \cdot 5} = \sqrt{9} \cdot \sqrt{5} = 3\sqrt{5}$

7. $\sqrt{48} = \sqrt{16 \cdot 3} = \sqrt{16} \cdot \sqrt{3} = 4\sqrt{3}$

9. $\sqrt{200} = \sqrt{100 \cdot 2} = \sqrt{100} \cdot \sqrt{2} = 10\sqrt{2}$

11. $\sqrt{147} = \sqrt{49 \cdot 3} = \sqrt{49} \cdot \sqrt{3} = 7\sqrt{3}$

13. $3\sqrt{12} = 3\sqrt{4 \cdot 3} = 3 \cdot \sqrt{4} \cdot \sqrt{3} = 3 \cdot 2\sqrt{3} = 6\sqrt{3}$

15. $\sqrt{5x^2} = \sqrt{5} \cdot \sqrt{x^2} = x\sqrt{5}$

17. $\sqrt{3y^4} = \sqrt{3(y^2)^2} = \sqrt{3} \cdot \sqrt{(y^2)^2} = y^2\sqrt{3}$

19. $\sqrt{2r^3} = \sqrt{2r \cdot r^2} = \sqrt{2r} \cdot \sqrt{r^2} = r\sqrt{2r}$

21. $\sqrt{27b^2} = \sqrt{9b^2 \cdot 3} = \sqrt{9b^2} \cdot \sqrt{3} = 3b\sqrt{3}$

23. $\sqrt{24x^4} = \sqrt{4x^4 \cdot 6} = \sqrt{4x^4} \cdot \sqrt{6} = 2x^2\sqrt{6}$

25. $\sqrt{54a^5} = \sqrt{9a^4 \cdot 6a} = \sqrt{9a^4} \cdot \sqrt{6a} = 3a^2\sqrt{6a}$

27. $\sqrt{x^3y^2} = \sqrt{x^2y^2x} = \sqrt{x^2y^2} \cdot \sqrt{x} = xy\sqrt{x}$

29. $\sqrt{\dfrac{4}{25}} = \dfrac{\sqrt{4}}{\sqrt{25}} = \dfrac{2}{5}$

31. $\sqrt{\dfrac{9}{16}} = \dfrac{\sqrt{9}}{\sqrt{16}} = \dfrac{3}{4}$

33. $\sqrt{\dfrac{3}{4}} = \dfrac{\sqrt{3}}{\sqrt{4}} = \dfrac{\sqrt{3}}{2}$

35. $\sqrt{\dfrac{5}{36}} = \dfrac{\sqrt{5}}{\sqrt{36}} = \dfrac{\sqrt{5}}{6}$

37. $\sqrt{\dfrac{8a^2}{25}} = \dfrac{\sqrt{8a^2}}{\sqrt{25}}$

$= \dfrac{\sqrt{4a^2 \cdot 2}}{5}$

$= \dfrac{\sqrt{4a^2} \cdot \sqrt{2}}{5}$

$= \dfrac{2a\sqrt{2}}{5}$

39. $\sqrt{\dfrac{1}{5}} = \dfrac{\sqrt{1}}{\sqrt{5}} = \dfrac{1 \cdot \sqrt{5}}{\sqrt{5} \cdot \sqrt{5}} = \dfrac{\sqrt{5}}{5}$

41. $\sqrt{\dfrac{3}{2}} = \dfrac{\sqrt{3}}{\sqrt{2}} = \dfrac{\sqrt{3} \cdot \sqrt{2}}{\sqrt{2} \cdot \sqrt{2}} = \dfrac{\sqrt{6}}{2}$

43. $\sqrt{\dfrac{3a}{5}} = \dfrac{\sqrt{3a}}{\sqrt{5}} = \dfrac{\sqrt{3a} \cdot \sqrt{5}}{\sqrt{5} \cdot \sqrt{5}} = \dfrac{\sqrt{15a}}{5}$

45. $\sqrt{\dfrac{2x^2}{3}} = \dfrac{\sqrt{2x^2}}{\sqrt{3}}$

$= \dfrac{\sqrt{2x^2} \cdot \sqrt{3}}{\sqrt{3} \cdot \sqrt{3}}$

$= \dfrac{\sqrt{x^2} \cdot \sqrt{2} \cdot \sqrt{3}}{3}$

$= \dfrac{x\sqrt{6}}{3}$

47. $\sqrt{\dfrac{8s^3}{7}} = \dfrac{\sqrt{8s^3}}{\sqrt{7}}$

$= \dfrac{\sqrt{4s^2 \cdot 2s} \cdot \sqrt{7}}{\sqrt{7} \cdot \sqrt{7}}$

$= \dfrac{\sqrt{4s^2} \cdot \sqrt{2s} \cdot \sqrt{7}}{7}$

$= \dfrac{2s\sqrt{14s}}{7}$

49. $\sqrt{10mn}$ is in simplest form.

51. $\dfrac{\sqrt{98x^2y}}{7x} = \dfrac{\sqrt{49x^2 \cdot 2y}}{7x} = \dfrac{7x\sqrt{2y}}{7x} = \sqrt{2y}$

Remove the perfect square factors from the radical and reduce the fraction.

53. Writing exercise

Exercises 9.3

1. $2\sqrt{2} + 4\sqrt{2} = (2+4)\sqrt{2} = 6\sqrt{2}$

3. $11\sqrt{7} - 4\sqrt{7} = (11-4)\sqrt{7} = 7\sqrt{7}$

5. $5\sqrt{7} + 3\sqrt{6}$; Cannot be simplified

7. $2\sqrt{3} - 5\sqrt{3} = (2-5)\sqrt{3} = -3\sqrt{3}$

9. $2\sqrt{3x} + 5\sqrt{3x} = (2+5)\sqrt{3x} = 7\sqrt{3x}$

11. $2\sqrt{3} + \sqrt{3} + 3\sqrt{3} = (2+1+3)\sqrt{3} = 6\sqrt{3}$

13. $5\sqrt{7} - 2\sqrt{7} + \sqrt{7} = (5-2+1)\sqrt{7} = 4\sqrt{7}$

15. $2\sqrt{5x} + 5\sqrt{5x} - 2\sqrt{5x} = (2+5-2)\sqrt{5x}$

$= 5\sqrt{5x}$

17. $2\sqrt{3} + \sqrt{12} = 2\sqrt{3} + \sqrt{4 \cdot 3}$

$= 2\sqrt{3} + \sqrt{4} \cdot \sqrt{3}$

$= 2\sqrt{3} + 2\sqrt{3}$

$= (2+2)\sqrt{3}$

$= 4\sqrt{3}$

19. $\sqrt{20} - \sqrt{5} = \sqrt{4 \cdot 5} - \sqrt{5}$

$= \sqrt{4} \cdot \sqrt{5} - \sqrt{5}$

$= 2\sqrt{5} - \sqrt{5}$

$= (2-1)\sqrt{5}$

$= \sqrt{5}$

21. $2\sqrt{6} - \sqrt{54} = 2\sqrt{6} - \sqrt{9 \cdot 6}$

$= 2\sqrt{6} - \sqrt{9} \cdot \sqrt{6}$

$= 2\sqrt{6} - 3\sqrt{6}$

$= (2-3)\sqrt{6}$

$= -\sqrt{6}$

23. $\sqrt{72} + \sqrt{50} = \sqrt{36 \cdot 2} + \sqrt{25 \cdot 2}$

$= \sqrt{36} \cdot \sqrt{2} + \sqrt{25} \cdot \sqrt{2}$

$= 6\sqrt{2} + 5\sqrt{2}$

$= (6+5)\sqrt{2}$

$= 11\sqrt{2}$

25. $3\sqrt{12} - \sqrt{48} = 3\sqrt{4 \cdot 3} - \sqrt{16 \cdot 3}$

$= 3 \cdot 2\sqrt{3} - 4\sqrt{3}$

$= 6\sqrt{3} - 4\sqrt{3}$

$= (6-4)\sqrt{3}$

$= 2\sqrt{3}$

27. $2\sqrt{45} - 2\sqrt{20} = 2\sqrt{9 \cdot 5} - 2\sqrt{4 \cdot 5}$

$= 2 \cdot 3\sqrt{5} - 2 \cdot 2\sqrt{5}$

$= 6\sqrt{5} - 4\sqrt{5}$

$= (6-4)\sqrt{5}$

$= 2\sqrt{5}$

29. $\sqrt{12} + \sqrt{27} - \sqrt{3} = \sqrt{4 \cdot 3} + \sqrt{9 \cdot 3} - \sqrt{3}$
$$= 2\sqrt{3} + 3\sqrt{3} - \sqrt{3}$$
$$= (2 + 3 - 1)\sqrt{3}$$
$$= 4\sqrt{3}$$

31. $3\sqrt{24} - \sqrt{54} + \sqrt{6} = 3 \cdot 2\sqrt{6} - 3\sqrt{6} + \sqrt{6}$
$$= 6\sqrt{6} - 3\sqrt{6} + \sqrt{6}$$
$$= (6 - 3 + 1)\sqrt{6}$$
$$= 4\sqrt{6}$$

33. $2\sqrt{50} + 3\sqrt{18} - \sqrt{32} = 2 \cdot 5\sqrt{2} + 3 \cdot 3\sqrt{2} - 4\sqrt{2}$
$$= 10\sqrt{2} + 9\sqrt{2} - 4\sqrt{2}$$
$$= (10 + 9 - 4)\sqrt{2}$$
$$= 15\sqrt{2}$$

35. $a\sqrt{27} - 2\sqrt{3a^2} = a\sqrt{9} \cdot \sqrt{3} - 2\sqrt{a^2} \cdot \sqrt{3}$
$$= 3a\sqrt{3} - 2a\sqrt{3}$$
$$= (3a - 2a)\sqrt{3}$$
$$= a\sqrt{3}$$

37. $5\sqrt{3x^3} + 2\sqrt{27x} = 5\sqrt{x^2} \cdot \sqrt{3x} + 2\sqrt{9} \cdot \sqrt{3x}$
$$= 5x\sqrt{3x} + 2 \cdot 3\sqrt{3x}$$
$$= (5x + 6)\sqrt{3x}$$

39. $\sqrt{3} - \sqrt{2} \approx 0.32$

41. $\sqrt{5} + \sqrt{3} \approx 3.97$

43. $4\sqrt{3} - 7\sqrt{5} \approx -8.72$

45. $5\sqrt{7} + 8\sqrt{13} \approx 42.07$

47. $P = 2L + 2W$
$$= 2\sqrt{36} + 2\sqrt{49}$$
$$= 2 \cdot 6 + 2 \cdot 7$$
$$= 12 + 14$$
$$= 26$$

49. $P = a + b + c$
$$= \left(\sqrt{3} - \sqrt{2}\right) + \left(\sqrt{3} + \sqrt{2}\right) + 3$$
$$= \sqrt{3} + \sqrt{3} + 3$$
$$= 2\sqrt{3} + 3$$

Exercises 9.4

1. $\sqrt{7} \cdot \sqrt{5} = \sqrt{7 \cdot 5} = \sqrt{35}$

3. $\sqrt{5} \cdot \sqrt{11} = \sqrt{5 \cdot 11} = \sqrt{55}$

5. $\sqrt{3} \cdot \sqrt{10m} = \sqrt{3 \cdot 10m} = \sqrt{30m}$

7. $\sqrt{2x} \cdot \sqrt{15} = \sqrt{2x \cdot 15} = \sqrt{30x}$

9. $\sqrt{3} \cdot \sqrt{7} \cdot \sqrt{2} = \sqrt{3 \cdot 7 \cdot 2} = \sqrt{42}$

11. $\sqrt{3} \cdot \sqrt{12} = \sqrt{3 \cdot 12} = \sqrt{36} = 6$

13. $\sqrt{10} \cdot \sqrt{10} = \sqrt{10 \cdot 10} = \sqrt{100} = 10$

15. $\sqrt{18} \cdot \sqrt{6} = \sqrt{18 \cdot 6}$
$$= \sqrt{108}$$
$$= \sqrt{36 \cdot 3}$$
$$= \sqrt{36} \cdot \sqrt{3}$$
$$= 6\sqrt{3}$$

17. $\sqrt{2x} \cdot \sqrt{6x} = \sqrt{2x \cdot 6x}$
$$= \sqrt{12x^2}$$
$$= \sqrt{4x^2 \cdot 3}$$
$$= 2x\sqrt{3}$$

19. $2\sqrt{3} \cdot \sqrt{7} = 2\sqrt{3 \cdot 7} = 2\sqrt{21}$

21. $\left(3\sqrt{3}\right)\left(5\sqrt{7}\right) = (3 \cdot 5)\sqrt{3 \cdot 7} = 15\sqrt{21}$

23. $\left(3\sqrt{5}\right)\left(2\sqrt{10}\right) = (3 \cdot 2)\sqrt{5 \cdot 10}$
$$= 6\sqrt{50}$$
$$= 6 \cdot 5\sqrt{2}$$
$$= 30\sqrt{2}$$

25. $\sqrt{5}\left(\sqrt{2} + \sqrt{5}\right) = \sqrt{5} \cdot \sqrt{2} + \sqrt{5} \cdot \sqrt{5}$
$$= \sqrt{10} + \sqrt{25}$$
$$= \sqrt{10} + 5$$

27. $\sqrt{3}\left(2\sqrt{5} - 3\sqrt{3}\right)$
$$= 2\sqrt{5} \cdot \sqrt{3} - 3\sqrt{3} \cdot \sqrt{3}$$
$$= 2\sqrt{15} - 3\sqrt{9}$$
$$= 2\sqrt{15} - 3 \cdot 3$$
$$= 2\sqrt{15} - 9$$

29. $\left(\sqrt{3}+5\right)\left(\sqrt{3}+3\right)=\sqrt{9}+3\sqrt{3}+5\sqrt{3}+15$

$\qquad\qquad 3+(3+5)\sqrt{3}+15$

$\qquad\qquad =18+8\sqrt{3}$

31. $\left(\sqrt{5}-1\right)\left(\sqrt{5}+3\right)=\sqrt{25}+3\sqrt{5}-\sqrt{5}-3$

$\qquad\qquad 5+(3-1)\sqrt{5}-3$

$\qquad\qquad =2+2\sqrt{5}$

33. $\left(\sqrt{5}-2\right)\left(\sqrt{5}+2\right)=\sqrt{25}-4=5-4=1$

35. $\left(\sqrt{10}+5\right)\left(\sqrt{10}-5\right)=\sqrt{100}-25$

$\qquad\qquad\qquad =10-25$

$\qquad\qquad\qquad =-15$

37. $\left(\sqrt{x}+3\right)\left(\sqrt{x}-3\right)=\sqrt{x^2}-9=x-9$

39. $\left(\sqrt{3}+2\right)^2=\sqrt{9}+2\cdot\sqrt{3}\cdot 2+4=7+4\sqrt{3}$

41. $\left(\sqrt{y}-5\right)^2=\sqrt{y^2}-2\cdot\sqrt{y}\cdot 5+25$

$\qquad\qquad =y-10\sqrt{y}+25$

43. $\dfrac{\sqrt{98}}{\sqrt{2}}=\sqrt{\dfrac{98}{2}}=\sqrt{49}=7$

45. $\dfrac{\sqrt{72a^2}}{\sqrt{2}}=\sqrt{\dfrac{72a^2}{2}}=\sqrt{36a^2}=6a$

47. $\dfrac{4+\sqrt{48}}{4}=\dfrac{4+\sqrt{16\cdot 3}}{4}$

$\qquad\quad =\dfrac{4+\sqrt{16}\cdot\sqrt{3}}{4}$

$\qquad\quad =\dfrac{4+4\sqrt{3}}{4}$

$\qquad\quad =\dfrac{4\left(1+\sqrt{3}\right)}{4}$

$\qquad\quad =1+\sqrt{3}$

49. $\dfrac{5+\sqrt{175}}{5}=\dfrac{5+\sqrt{25\cdot 7}}{5}$

$\qquad\quad =\dfrac{5+\sqrt{25}\cdot\sqrt{7}}{5}$

$\qquad\quad =\dfrac{5+5\sqrt{7}}{5}$

$\qquad\quad =\dfrac{5\left(1+\sqrt{7}\right)}{5}$

$\qquad\quad =1+\sqrt{7}$

51. $\dfrac{-8-\sqrt{512}}{4}=\dfrac{-8-\sqrt{256\cdot 2}}{4}$

$\qquad\quad =\dfrac{-8-\sqrt{256}\cdot\sqrt{2}}{4}$

$\qquad\quad =\dfrac{-8-16\sqrt{2}}{4}$

$\qquad\quad =\dfrac{4\left(-2-4\sqrt{2}\right)}{4}$

$\qquad\quad =-2-4\sqrt{2}$

53. $\dfrac{6+\sqrt{18}}{3}=\dfrac{6+\sqrt{9\cdot 2}}{3}$

$\qquad\quad =\dfrac{6+\sqrt{9}\cdot\sqrt{2}}{3}$

$\qquad\quad =\dfrac{6+3\sqrt{2}}{3}$

$\qquad\quad =\dfrac{3\left(2+\sqrt{2}\right)}{3}$

$\qquad\quad =2+\sqrt{2}$

55. $\dfrac{15-\sqrt{75}}{5}=\dfrac{15-\sqrt{25\cdot 3}}{5}$

$\qquad\quad =\dfrac{15-\sqrt{25}\cdot\sqrt{3}}{5}$

$\qquad\quad =\dfrac{15-5\sqrt{3}}{5}$

$\qquad\quad =\dfrac{5\left(3-\sqrt{3}\right)}{5}$

$\qquad\quad =3-\sqrt{3}$

57. $A=l\cdot w$

$\qquad =\sqrt{3}\cdot\sqrt{11}$

$\qquad =\sqrt{33}$

59. Writing exercise

61. Group exercise

Exercises 9.5

1. $x^2 = 12^2 + 9^2$
$x^2 = 144 + 81$
$x^2 = 225$
$x = 15$ Use the positive square root.

3. $x^2 + 8^2 = 17^2$
$x^2 + 64 = 289$
$x^2 = 225$
$x = 15$ Use the positive square root.

5. $x^2 + 5^2 = 7^2$
$x^2 + 25 = 49$
$x^2 = 24$
$x = 2\sqrt{6}$ Use the positive square root.

7.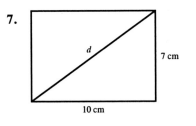

$d^2 = 10^2 + 7^2$
$d^2 = 100 + 49$
$d^2 = 149$
$d \approx 12.207$ cm
Find the positive square root using a calculator.

9.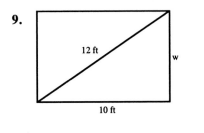

$10^2 + w^2 = 12^2$
$100 + w^2 = 144$
$w^2 = 44$
$w \approx 6.633$ ft
Find the positive square root using a calculator.

11.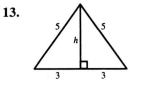

$l^2 = 20^2 + 8^2$
$l^2 = 400 + 64$
$l^2 = 464$
$l \approx 21.541$ ft
Find the positive square root using a calculator.

13.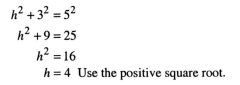

$h^2 + 3^2 = 5^2$
$h^2 + 9 = 25$
$h^2 = 16$
$h = 4$ Use the positive square root.

15.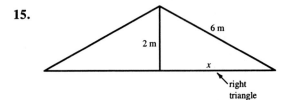

$x^2 + 2^2 = 6^2$
$x^2 + 4 = 36$
$x^2 = 32$
$x = 4\sqrt{2}$
Since the insulation must cover the full width of the attic, each piece should be $2x$ long.
$2\left(4\sqrt{2}\right) = 8\sqrt{2} \approx 11.3$ m

17. Let x = length of adjusted base

$$x^2 + (3+0.5)^2 = 5^2$$
$$x^2 + 3.5^2 = 5^2$$
$$x^2 + 12.25 = 25$$
$$x^2 = 12.75$$
$$x \approx 3.6 \text{ m}$$

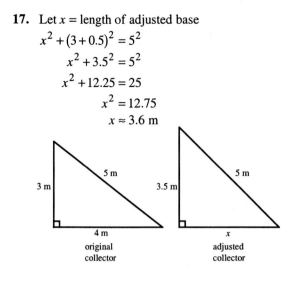

original collector

adjusted collector

19. (2, 0) and (–4, 0)

$$d = \sqrt{(-4-2)^2 + (0-0)^2}$$
$$= \sqrt{(-6)^2 + (0)^2}$$
$$= \sqrt{36}$$
$$= 6$$

21. (0, –2) and (0, –9)

$$d = \sqrt{(0-0)^2 + [-9-(-2)]^2}$$
$$= \sqrt{(0)^2 + (-7)^2}$$
$$= \sqrt{49}$$
$$= 7$$

23. (2, 5) and (5, 2)

$$d = \sqrt{(5-2)^2 + (2-5)^2}$$
$$= \sqrt{(3)^2 + (-3)^2}$$
$$= \sqrt{9+9}$$
$$= \sqrt{18}$$
$$= 3\sqrt{2}$$

25. (5, 1) and (3, 8)

$$d = \sqrt{(3-5)^2 + (8-1)^2}$$
$$= \sqrt{(-2)^2 + (7)^2}$$
$$= \sqrt{4+49}$$
$$= \sqrt{53}$$

27. (–2, 8) and (1, 5)

$$d = \sqrt{[1-(-2)]^2 + (5-8)^2}$$
$$= \sqrt{(3)^2 + (-3)^2}$$
$$= \sqrt{9+9}$$
$$= \sqrt{18}$$
$$= 3\sqrt{2}$$

29. (6, –1) and (2, 2)

$$d = \sqrt{(2-6)^2 + [2-(-1)]^2}$$
$$= \sqrt{(-4)^2 + (3)^2}$$
$$= \sqrt{16+9}$$
$$= \sqrt{25}$$
$$= 5$$

31. (–1, –1) and (2, 5)

$$d = \sqrt{[2-(-1)]^2 + [5-(-1)]^2}$$
$$= \sqrt{(3)^2 + (6)^2}$$
$$= \sqrt{9+36}$$
$$= \sqrt{45}$$
$$= 3\sqrt{5}$$

33. (–2, 9) and (–3, 3)

$$d = \sqrt{[-3-(-2)]^2 + (3-9)^2}$$
$$= \sqrt{(-1)^2 + (-6)^2}$$
$$= \sqrt{1+36}$$
$$= \sqrt{37}$$

35. (–1, –4) and (–3, 5)

$$d = \sqrt{[-3-(-1)]^2 + [5-(-4)]^2}$$
$$= \sqrt{(-2)^2 + (9)^2}$$
$$= \sqrt{4+81}$$
$$= \sqrt{85}$$

37. (–2, –4) and (–4, 1)

$$d = \sqrt{[-4-(-2)]^2 + [1-(-4)]^2}$$
$$= \sqrt{(-2)^2 + (5)^2}$$
$$= \sqrt{4+25}$$
$$= \sqrt{29}$$

39. $(-4, -2)$ and $(-1, -5)$

$$d = \sqrt{[-1-(-4)]^2 + [-5-(-2)]^2}$$
$$= \sqrt{(3)^2 + (-3)^2}$$
$$= \sqrt{9+9}$$
$$= \sqrt{18}$$
$$= 3\sqrt{2}$$

41. $(-2, 0)$ and $(-4, -1)$

$$d = \sqrt{[-4-(-2)]^2 + (-1-0)^2}$$
$$= \sqrt{(-2)^2 + (-1)^2}$$
$$= \sqrt{4+1}$$
$$= \sqrt{5}$$

43. $A = (-3, 0)$, $B = (2, 3)$, and $C = (1, -1)$

$$d(A, B) = \sqrt{[2-(-3)]^2 + (3-0)^2} = \sqrt{25+9}$$
$$= \sqrt{34}$$
$$d(A, C) = \sqrt{(-3-1)^2 + [0-(-1)]^2} = \sqrt{4^2+1^2}$$
$$= \sqrt{16+1} = \sqrt{17}$$
$$d(B, C) = \sqrt{(2-1)^2 + [3-(-1)]^2} = \sqrt{1^2+4^2}$$
$$= \sqrt{1+16} = \sqrt{17}$$

Thus $d(A, C) = d(B, C)$.

45. Let x be the length of the shorter leg.

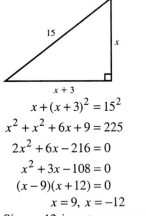

$$x + (x+3)^2 = 15^2$$
$$x^2 + x^2 + 6x + 9 = 225$$
$$2x^2 + 6x - 216 = 0$$
$$x^2 + 3x - 108 = 0$$
$$(x-9)(x+12) = 0$$
$$x = 9, \ x = -12$$

Since -12 is not a reasonable answer, the lengths of the sides of the triangle are 9 in. and 12 in.

47. $(1, 1)$ and $(5, 2)$

$$d = \sqrt{(5-1)^2 + (2-1)^2}$$
$$= \sqrt{4^2+1^2}$$
$$= \sqrt{16+1}$$
$$= \sqrt{17}$$
$$\approx 4.12$$

49. $(1, 2)$ and $(4, -2)$

$$d = \sqrt{(4-1)^2 + [(-2)-2]^2}$$
$$= \sqrt{3^2 + (-4)^2}$$
$$= \sqrt{9+16}$$
$$= \sqrt{25}$$
$$= 5$$

51. The coordinates of the vertices of the figure are $(0, -1)$, $(2, 2)$, $(5, 0)$, and $(3, -3)$. First, calculate the slopes of the segments connecting each pair of adjacent vertices.

$(0, -1)$ and $(2, 2)$: $m_1 = \dfrac{2-(-1)}{2-0} = \dfrac{3}{2}$

$(2, 2)$ and $(5, 0)$: $m_2 = \dfrac{0-2}{5-2} = -\dfrac{2}{3}$

$(5, 0)$ and $(3, -3)$: $m_3 = \dfrac{-3-0}{3-5} = \dfrac{-3}{-2} = \dfrac{3}{2}$

$(3, -3)$ and $(0, -1)$: $m_4 = \dfrac{-1-(-3)}{0-3} = \dfrac{2}{-3} = -\dfrac{2}{3}$

Since $m_1 \cdot m_2 = m_2 \cdot m_3 = m_3 \cdot m_4 = m_4 \cdot m_1 = -1$, each pair of adjacent sides is perpendicular. Therefore, all the angles in the figure are right angles.

Next, find the lengths of the sides of the figure.

$$d_1 = \sqrt{(2-0)^2 + [2-(-1)]^2}$$
$$= \sqrt{2^2 + 3^2}$$
$$= \sqrt{4+9}$$
$$= \sqrt{13}$$
$$d_2 = \sqrt{(5-2)^2 + (0-2)^2}$$
$$= \sqrt{3^2 + (-2)^2}$$
$$= \sqrt{9+4}$$
$$= \sqrt{13}$$
$$d_3 = \sqrt{(3-5)^2 + [(-3)-0]^2}$$
$$= \sqrt{(-2)^2 + (-3)^2}$$
$$= \sqrt{4+9}$$
$$= \sqrt{13}$$
$$d_4 = \sqrt{(0-3)^2 + [(-1)-(-3)]^2}$$
$$= \sqrt{(-3)^2 + 2^2}$$
$$= \sqrt{9+4}$$
$$= \sqrt{13}$$

Since all the sides of the figure have a length of $\sqrt{13}$ and all of the angles are right angles, the figure is a square.

$$\text{Area} = d^2$$
$$= \left(\sqrt{13}\right)^2$$
$$= 13$$

53. Challenge exercise

Summary Exercises for Chapter 9

1. $\sqrt{81} = \sqrt{9^2} = 9$

3. $\sqrt{-49}$ is not a real number, since $-49 < 0$.

5. $\sqrt[3]{-64} = \sqrt[3]{(-4)^3} = -4$

7. $\sqrt[4]{-81}$ is not a real number, since $-81 < 0$.

9. $\sqrt{45} = \sqrt{9 \cdot 5} = \sqrt{9} \cdot \sqrt{5} = 3\sqrt{5}$

11. $\sqrt{20x^4} = \sqrt{4x^4 \cdot 5} = \sqrt{4x^4} \cdot \sqrt{5} = 2x^2\sqrt{5}$

13. $\sqrt{200b^3} = \sqrt{100b^2 \cdot 2b}$
$$= \sqrt{100b^2} \cdot \sqrt{2b}$$
$$= 10b\sqrt{2b}$$

15. $\sqrt{108a^2b^5} = \sqrt{36a^2b^4 \cdot 3b}$
$$= \sqrt{36a^2b^4} \cdot \sqrt{3b}$$
$$= 6ab^2\sqrt{3b}$$

17. $\sqrt{\dfrac{18x^2}{25}} = \dfrac{\sqrt{18x^2}}{\sqrt{25}}$
$$= \dfrac{\sqrt{9x^2 \cdot 2}}{5}$$
$$= \dfrac{\sqrt{9x^2} \cdot \sqrt{2}}{5}$$
$$= \dfrac{3x\sqrt{2}}{5}$$

19. $\sqrt{\dfrac{3}{7}} = \dfrac{\sqrt{3}}{\sqrt{7}} = \dfrac{\sqrt{3} \cdot \sqrt{7}}{\sqrt{7} \cdot \sqrt{7}} = \dfrac{\sqrt{21}}{7}$

21. $\sqrt{\dfrac{8x^2}{7}} = \dfrac{\sqrt{8x^2}}{\sqrt{7}}$
$$= \dfrac{\sqrt{8x^2} \cdot \sqrt{7}}{\sqrt{7} \cdot \sqrt{7}}$$
$$= \dfrac{\sqrt{56x^2}}{7}$$
$$= \dfrac{\sqrt{4x^2 \cdot 14}}{7}$$
$$= \dfrac{2x\sqrt{14}}{7}$$

23. $9\sqrt{5} - 3\sqrt{5} = (9-3)\sqrt{5} = 6\sqrt{5}$

25. $3\sqrt{3a} - \sqrt{3a} = (3-1)\sqrt{3a} = 2\sqrt{3a}$

27. $5\sqrt{3} + \sqrt{12} = 5\sqrt{3} + \sqrt{4 \cdot 3}$
$$= 5\sqrt{3} + 2\sqrt{3}$$
$$= (5+2)\sqrt{3}$$
$$= 7\sqrt{3}$$

29. $\sqrt{32} - \sqrt{18} = \sqrt{16 \cdot 2} - \sqrt{9 \cdot 2}$
$$= 4\sqrt{2} - 3\sqrt{2}$$
$$= (4-3)\sqrt{2}$$
$$= \sqrt{2}$$

31. $\sqrt{8} + 2\sqrt{27} - \sqrt{75} = \sqrt{4 \cdot 2} + 2\sqrt{9 \cdot 3} - \sqrt{25 \cdot 3}$
$$= 2\sqrt{2} + 2 \cdot 3\sqrt{3} - 5\sqrt{3}$$
$$= 2\sqrt{2} + 6\sqrt{3} - 5\sqrt{3}$$
$$= 2\sqrt{2} + (6 - 5)\sqrt{3}$$
$$= 2\sqrt{2} + \sqrt{3}$$

33. $\sqrt{6} \cdot \sqrt{5} = \sqrt{6 \cdot 5} = \sqrt{30}$

35. $\sqrt{3x} \cdot \sqrt{2} = \sqrt{3x \cdot 2} = \sqrt{6x}$

37. $\sqrt{5a} \cdot \sqrt{10a} = \sqrt{5a \cdot 10a}$
$$= \sqrt{50a^2}$$
$$= \sqrt{25a^2 \cdot 2}$$
$$= 5a\sqrt{2}$$

39. $\sqrt{7}\left(2\sqrt{3} - 3\sqrt{7}\right) = 2\sqrt{21} - 3\sqrt{49}$
$$= 2\sqrt{21} - 3 \cdot 7$$
$$= 2\sqrt{21} - 21$$

41. $\left(\sqrt{15} - 3\right)\left(\sqrt{15} + 3\right) = \sqrt{15^2} - 9 = 15 - 9 = 6$

43. $\dfrac{\sqrt{7x^3}}{\sqrt{3}} = \dfrac{\sqrt{7x^3} \cdot \sqrt{3}}{\sqrt{3} \cdot \sqrt{3}} = \dfrac{\sqrt{21x^3}}{3} = \dfrac{x\sqrt{21x}}{3}$

45. $x^2 = 8^2 + 6^2$
$x^2 = 64 + 36$
$x^2 = 100$
$x = 10$ Use the positive square root.

47. $x^2 + 8^2 = 17^2$
$x^2 + 64 = 289$
$x^2 = 225$
$x = 15$ Use the positive square root.

49. $x^2 + 10^2 = 15^2$
$x^2 + 100 = 225$
$x^2 = 125$
$x = 5\sqrt{5}$ Use the positive square root.

51.

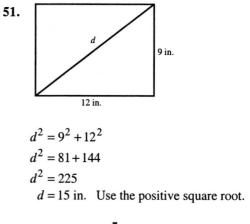

$d^2 = 9^2 + 12^2$
$d^2 = 81 + 144$
$d^2 = 225$
$d = 15$ in. Use the positive square root.

53.

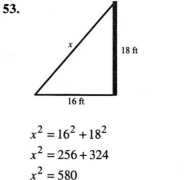

$x^2 = 16^2 + 18^2$
$x^2 = 256 + 324$
$x^2 = 580$
$x \approx 24.1$ ft Find the positive square
root using a calculator

55. $d = \sqrt{[-7 - (-3)]^2 + (2 - 2)^2}$
$$= \sqrt{(-4)^2 + 0^2}$$
$$= \sqrt{16}$$
$$= 4$$

57. $d = \sqrt{[-5 - (-2)]^2 + (-1 - 7)^2}$
$$= \sqrt{(-3)^2 + (-8)^2}$$
$$= \sqrt{9 + 64}$$
$$= \sqrt{73}$$

59. $d = \sqrt{[-2 - (-3)]^2 + (-5 - 4)^2}$
$$= \sqrt{1^2 + 9^2}$$
$$= \sqrt{1 + 81}$$
$$= \sqrt{82}$$

Chapter 10   Quadratic Equations

Exercises 10.1

1. $x^2 = 5$
$\quad x = \pm\sqrt{5}$

3. $x^2 = 33$
$\quad x = \pm\sqrt{33}$

5. $x^2 - 7 = 0$
$\quad x^2 = 7$
$\quad\quad x = \pm\sqrt{7}$

7. $x^2 - 20 = 0$
$\quad x^2 = 20$
$\quad\quad x = \pm 2\sqrt{5}$

9. $x^2 = 40$
$\quad x = \pm 2\sqrt{10}$

11. $x^2 + 3 = 12$
$\quad x^2 = 9$
$\quad\quad x = \pm 3$

13. $x^2 + 5 = 8$
$\quad x^2 = 3$
$\quad\quad x = \pm\sqrt{3}$

15. $x^2 - 2 = 16$
$\quad x^2 = 18$
$\quad\quad x = \pm 3\sqrt{2}$

17. $9x^2 = 25$
$\quad x^2 = \dfrac{25}{9}$
$\quad\quad x = \pm\dfrac{5}{3}$

19. $49x^2 = 11$
$\quad x^2 = \dfrac{11}{49}$
$\quad\quad x = \dfrac{\pm\sqrt{11}}{7}$

21. $4x^2 = 7$
$\quad x^2 = \dfrac{7}{4}$
$\quad\quad x = \dfrac{\pm\sqrt{7}}{2}$

23. $(x-1)^2 = 5$
$\quad x - 1 = \pm\sqrt{5}$
$\quad\quad x = 1 \pm\sqrt{5}$

25. $(x+1)^2 = 12$
$\quad x + 1 = \pm 2\sqrt{3}$
$\quad\quad x = -1 \pm 2\sqrt{3}$

27. $(x-3)^2 = 24$
$\quad x - 3 = \pm 2\sqrt{6}$
$\quad\quad x = 3 \pm 2\sqrt{6}$

29. $(x+5)^2 = 25$
$\quad x + 5 = \pm 5$
$\quad\quad x = -5 \pm 5$
$\quad\quad x = -10,\ 0$

31. $3(x-5)^2 = 7$
$\quad (x-5)^2 = \dfrac{7}{3}$
$\quad\quad x - 5 = \pm\sqrt{\dfrac{7}{3}}$
$\quad\quad x - 5 = \dfrac{\pm\sqrt{21}}{3}$
$\quad\quad x = 5 \pm \dfrac{\sqrt{21}}{3}$
$\quad\quad x = \dfrac{15 \pm \sqrt{21}}{3}$

33. $4(x+5)^2 = 9$
$\quad (x+5)^2 = \dfrac{9}{4}$
$\quad\quad x + 5 = \dfrac{\pm 3}{2}$
$\quad\quad x = -5 \pm \dfrac{3}{2}$
$\quad\quad x = \dfrac{-13}{2},\ \dfrac{-7}{2}$

35. $-2(x+2)^2 = -6$
$$(x+2)^2 = 3$$
$$x+2 = \pm\sqrt{3}$$
$$x = -2 \pm \sqrt{3}$$

37. $-4(x-2)^2 = -5$
$$(x-2)^2 = \frac{5}{4}$$
$$x-2 = \frac{\pm\sqrt{5}}{2}$$
$$x = 2 \pm \frac{\sqrt{5}}{2}$$
$$x = \frac{4 \pm \sqrt{5}}{2}$$

39. $(5x-2)^2 = 8$
$$5x-2 = \pm\, 2\sqrt{2}$$
$$5x = 2 \pm 2\sqrt{2}$$
$$x = \frac{2}{5} \pm \frac{2\sqrt{2}}{5}$$
$$x = \frac{2 \pm 2\sqrt{2}}{5}$$

41. $x^2 - 2x + 1 = 7$
$$(x-1)^2 = 7$$
$$x-1 = \pm\sqrt{7}$$
$$x = 1 \pm \sqrt{7}$$

43. $(2x+11)^2 + 9 = 0$
$$(2x+11)^2 = -9$$
$$2x+11 = \pm\sqrt{-9}$$
No real number solution.

45. Let x be the number.
$$x^2 - 2 = -x$$
$$x^2 + x - 2 = 0$$
$$(x+2)(x-1) = 0$$
$$x+2 = 0 \quad \text{or} \quad x-1 = 0$$
$$x = -2 \quad \text{or} \quad x = 1$$
The number is -2 or 1.

47.
$$R = x\left(5 - \frac{1}{10}x\right)$$
$$60 = 5x - \frac{1}{10}x^2$$
$$x^2 - 50x + 600 = 0$$
$$(x-20)(x-30) = 0$$
$$x - 20 = 0 \quad \text{or} \quad x - 30 = 0$$
$$x = 20 \quad \text{or} \quad x = 30$$
20 or 30 units must be sold to obtain a revenue of $60.

49. a.

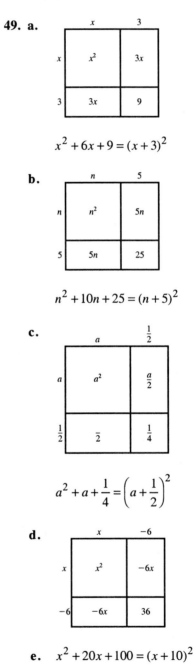

$$x^2 + 6x + 9 = (x+3)^2$$

b.

$$n^2 + 10n + 25 = (n+5)^2$$

c.

$$a^2 + a + \frac{1}{4} = \left(a + \frac{1}{2}\right)^2$$

d.

e. $x^2 + 20x + 100 = (x+10)^2$

f. $n^2 - 16n + 64 = (n-8)^2$

Exercises 10.2

1. $x^2 - 14x + 49 = (x-7)^2$
Yes

3. $x^2 - 18x - 81$
No, since the third term is negative

5. $x^2 - 18x + 81 = (x-9)^2$
Yes

7. $x^2 + 6x; \left(\dfrac{6}{2}\right)^2 = 9$

9. $x^2 - 10; \left(\dfrac{-10}{2}\right)^2 = 25$

11. $x^2 + 9x; \left(\dfrac{9}{2}\right)^2 = \dfrac{81}{4}$

13. $x^2 + 4x - 12 = 0$
$$x^2 + 4x = 12$$
$$x^2 + 4x + 4 = 12 + 4$$
$$(x+2)^2 = 16$$
$$x + 2 = \pm 4$$
$$x = -2 \pm 4$$
$$x = -6,\ 2$$

15. $x^2 - 2x - 5 = 0$
$$x^2 - 2x = 5$$
$$x^2 - 2x + 1 = 5 + 1$$
$$(x-1)^2 = 6$$
$$x - 1 = \pm\sqrt{6}$$
$$x = 1 \pm \sqrt{6}$$

17. $x^2 + 3x - 27 = 0$
$$x^2 + 3x = 27$$
$$x^2 + 3x + \dfrac{9}{4} = 27 + \dfrac{9}{4}$$
$$\left(x + \dfrac{3}{2}\right)^2 = \dfrac{117}{4}$$
$$x + \dfrac{3}{2} = \dfrac{\pm 3\sqrt{13}}{2}$$
$$x = \dfrac{-3 \pm 3\sqrt{13}}{2}$$

19. $x^2 + 6x - 1 = 0$
$$x^2 + 6x = 1$$
$$x^2 + 6x + 9 = 1 + 9$$
$$(x+3)^2 = 10$$
$$x + 3 = \pm\sqrt{10}$$
$$x = -3 \pm \sqrt{10}$$

21. $x^2 - 5x + 6 = 0$
$$x^2 - 5x = -6$$
$$x^2 - 5x + \dfrac{25}{4} = -6 + \dfrac{25}{4}$$
$$\left(x - \dfrac{5}{2}\right)^2 = \dfrac{1}{4}$$
$$x - \dfrac{5}{2} = \dfrac{\pm 1}{2}$$
$$x = \dfrac{5}{2} \pm \dfrac{1}{2}$$
$$x = 2,\ 3$$

23. $x^2 + 6x - 5 = 0$
$$x^2 + 6x = 5$$
$$x^2 + 6x + 9 = 5 + 9$$
$$(x+3)^2 = 14$$
$$x + 3 = \pm\sqrt{14}$$
$$x = -3 \pm \sqrt{14}$$

25. $x^2 = 9x + 5$
$$x^2 - 9x = 5$$
$$x^2 - 9x + \dfrac{81}{4} = 5 + \dfrac{81}{4}$$
$$\left(x - \dfrac{9}{2}\right)^2 = \dfrac{101}{4}$$
$$x - \dfrac{9}{2} = \dfrac{\pm\sqrt{101}}{2}$$
$$x = \dfrac{9 \pm \sqrt{101}}{2}$$

27. $2x^2 - 6x + 1 = 0$

$$2x^2 - 6x = -1$$

$$x^2 - 3x = \frac{-1}{2}$$

$$x^2 - 3x + \frac{9}{4} = \frac{-1}{2} + \frac{9}{4}$$

$$\left(x - \frac{3}{2}\right)^2 = \frac{7}{4}$$

$$x - \frac{3}{2} = \frac{\pm\sqrt{7}}{2}$$

$$x = \frac{3 \pm \sqrt{7}}{2}$$

29. $2x^2 - 4x + 1 = 0$

$$2x^2 - 4x = -1$$

$$x^2 - 2x = \frac{-1}{2}$$

$$x^2 - 2x + 1 = \frac{-1}{2} + 1$$

$$(x - 1)^2 = \frac{1}{2}$$

$$x - 1 = \pm\sqrt{\frac{1}{2}}$$

$$x - 1 = \frac{\pm\sqrt{2}}{2}$$

$$x = 1 \pm \frac{\sqrt{2}}{2}$$

$$x = \frac{2 \pm \sqrt{2}}{2}$$

31. $4x^2 - 2x - 1 = 0$

$$4x^2 - 2x = 1$$

$$x^2 - \frac{x}{2} = \frac{1}{4}$$

$$x^2 - \frac{x}{2} + \frac{1}{16} = \frac{1}{4} + \frac{1}{16}$$

$$\left(x - \frac{1}{4}\right)^2 = \frac{5}{16}$$

$$x - \frac{1}{4} = \frac{\pm\sqrt{5}}{4}$$

$$x = \frac{1 \pm \sqrt{5}}{4}$$

33. $3x^2 - 4x + 7x - 9 = 2x^2 + 5x - 4$

$$3x^2 + 3x - 9 = 2x^2 + 5x - 4$$

$$x^2 - 2x - 5 = 0$$

$$x^2 - 2x = 5$$

$$x^2 - 2x + 1 = 5 + 1$$

$$(x - 1)^2 = 6$$

$$x - 1 = \pm\sqrt{6}$$

$$x = 1 \pm \sqrt{6}$$

35. Let x be the number.

$$(x + 3)^2 = 9$$

$$x + 3 = \pm 3$$

$$x = -3 \pm 3$$

$$x = 0, \, -6$$

The number is 0 or –6.

37.

$$R = x\left(25 - \frac{1}{2}x\right)$$

$$294.50 = x\left(25 - \frac{1}{2}x\right)$$

$$294.50 = 25x - \frac{1}{2}x^2$$

$$x^2 - 50x + 589 = 0$$

$$x^2 - 50x = -589$$

$$x^2 - 50x + 625 = -589 + 625$$

$$(x - 25)^2 = 36$$

$$x - 25 = \pm 6$$

$$x = 25 \pm 6$$

$$x = 31, \, 19$$

The number of units sold was 19 or 31.

Exercises 10.3

1. $x^2 + 9x + 20 = 0$

$$x = \frac{-9 \pm \sqrt{9^2 - 4(1)(20)}}{2(1)}$$

$$x = \frac{-9 \pm 1}{2}$$

$$x = -5, \, -4$$

3. $x^2 - 4x + 3 = 0$

$$x = \frac{4 \pm \sqrt{(-4)^2 - 4(1)(3)}}{2(1)}$$

$$x = \frac{4 \pm 2}{2}$$

$$x = 1, \, 3$$

5. $3x^2 + 2x - 1 = 0$

$$x = \frac{-2 \pm \sqrt{2^2 - 4(3)(-1)}}{2(3)}$$

$$x = \frac{-2 \pm 4}{6}$$

$$x = -1, \ \frac{1}{3}$$

7. $x^2 + 5x = -4$

$x^2 + 5x + 4 = 0$

$$x = \frac{-5 \pm \sqrt{5^2 - 4(1)(4)}}{2(1)}$$

$$x = \frac{-5 \pm 3}{2}$$

$$x = -4, \ -1$$

9. $x^2 = 6x - 9$

$x^2 - 6x + 9 = 0$

$$x = \frac{6 \pm \sqrt{(-6)^2 - 4(1)(9)}}{2(1)}$$

$$x = \frac{6 \pm 0}{2}$$

$$x = 3$$

11. $2x^2 - 3x - 7 = 0$

$$x = \frac{3 \pm \sqrt{(-3)^2 - 4(2)(-7)}}{2(2)}$$

$$x = \frac{3 \pm \sqrt{65}}{4}$$

13. $x^2 + 2x - 4 = 0$

$$x = \frac{-2 \pm \sqrt{2^2 - 4(1)(-4)}}{2(1)}$$

$$x = \frac{-2 \pm \sqrt{20}}{2}$$

$$x = \frac{-2 \pm 2\sqrt{5}}{2}$$

$$x = -1 \pm \sqrt{5}$$

15. $2x^2 - 3x = 3$

$2x^2 - 3x - 3 = 0$

$$x = \frac{3 \pm \sqrt{(-3)^2 - 4(2)(-3)}}{2(2)}$$

$$x = \frac{3 \pm \sqrt{33}}{4}$$

17. $3x^2 - 2x = 6$

$3x^2 - 2x - 6 = 0$

$$x = \frac{2 \pm \sqrt{(-2)^2 - 4(3)(-6)}}{2(3)}$$

$$x = \frac{2 \pm \sqrt{76}}{6}$$

$$x = \frac{2 \pm 2\sqrt{19}}{6}$$

$$x = \frac{1 \pm \sqrt{19}}{3}$$

19. $3x^2 + 3x + 2 = 0$

$$x = \frac{-3 \pm \sqrt{3^2 - 4(3)(2)}}{2(3)}$$

$$x = \frac{-3 \pm \sqrt{-15}}{6}$$

No real number solutions

21. $5x^2 = 8x - 2$

$5x^2 - 8x + 2 = 0$

$$x = \frac{8 \pm \sqrt{(-8)^2 - 4(5)(2)}}{2(5)}$$

$$x = \frac{8 \pm \sqrt{24}}{10}$$

$$x = \frac{8 \pm 2\sqrt{6}}{10}$$

$$x = \frac{4 \pm \sqrt{6}}{5}$$

23. $2x^2 - 9 = 4x$

$2x^2 - 4x - 9 = 0$

$$x = \frac{4 \pm \sqrt{(-4)^2 - 4(2)(-9)}}{2(2)}$$

$$x = \frac{4 \pm \sqrt{88}}{4}$$

$$x = \frac{4 \pm 2\sqrt{22}}{4}$$

$$x = \frac{2 \pm \sqrt{22}}{2}$$

25.

$$3x - 5 = \frac{1}{x}$$
$$3x^2 - 5x = 1$$
$$3x^2 - 5x - 1 = 0$$
$$x = \frac{5 \pm \sqrt{(-5)^2 - 4(3)(-1)}}{2(3)}$$
$$x = \frac{5 \pm \sqrt{37}}{6}$$

27.

$$(x-2)(x+1) = 3$$
$$x^2 - x - 2 = 3$$
$$x^2 - x - 5 = 0$$
$$x = \frac{1 \pm \sqrt{(-1)^2 - 4(1)(-5)}}{2(1)}$$
$$x = \frac{1 \pm \sqrt{21}}{2}$$

29.

$$(x-1)^2 = 7$$
$$x - 1 = \pm\sqrt{7}$$
$$x = 1 \pm \sqrt{7}$$

31.

$$x^2 - 5x - 14 = 0$$
$$(x-7)(x+2) = 0$$
$$x - 7 = 0 \text{ or } x + 2 = 0$$
$$x = 7 \text{ or } x = -2$$

33.

$$6x^2 - 23x + 10 = 0$$
$$(2x-1)(3x-10) = 0$$
$$2x - 1 = 0 \text{ or } 3x - 10 = 0$$
$$x = \frac{1}{2} \text{ or } x = \frac{10}{3}$$

35.

$$2x^2 - 8x + 3 = 0$$
$$x = \frac{8 \pm \sqrt{(-8)^2 - 4(2)(3)}}{2(2)}$$
$$x = \frac{8 \pm \sqrt{40}}{4}$$
$$x = \frac{8 \pm 2\sqrt{10}}{4}$$
$$x = \frac{4 \pm \sqrt{10}}{2}$$

37.

$$x^2 - 9x - 4 = 6$$
$$x^2 - 9x - 10 = 0$$
$$(x-10)(x+1) = 0$$
$$x - 10 = 0 \text{ or } x + 1 = 0$$
$$x = 10 \text{ or } x = -1$$

39.

$$4x^2 - 8x + 3 = 5$$
$$4x^2 - 8x - 2 = 0$$
$$2(2x^2 - 4x - 1) = 0$$
$$2x^2 - 4x - 1 = 0$$
$$x = \frac{-(-4) \pm \sqrt{(-4)^2 - 4(2)(-1)}}{2(2)}$$
$$x = \frac{4 \pm \sqrt{24}}{4}$$
$$x = \frac{4 \pm 2\sqrt{6}}{4}$$
$$x = \frac{2 \pm \sqrt{6}}{2}$$

41.

$$\frac{3}{x} + \frac{5}{x^2} = 9$$
$$3x + 5 = 9x^2$$
$$9x^2 - 3x - 5 = 0$$
$$x = \frac{3 \pm \sqrt{(-3)^2 - 4(9)(-5)}}{2(9)}$$
$$x = \frac{3 \pm \sqrt{189}}{18}$$
$$x = \frac{3 \pm 3\sqrt{21}}{18}$$
$$x = \frac{1 \pm \sqrt{21}}{6}$$

43.

$$\frac{x}{x+1} + \frac{10x}{x^2 + 4x + 3} = \frac{15}{x+3}$$
$$x(x+3) + 10x = 15(x+1)$$
$$x^2 + 3x + 10x = 15x + 15$$
$$x^2 - 2x - 15 = 0$$
$$(x-5)(x+3) = 0$$
$$x - 5 = 0 \text{ or } x + 3 = 0$$
$$x = 5 \text{ or } \quad x = -3$$

Reject $x = -3$ since it gives division by zero.

$$x = 5$$

45.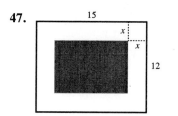

$$4x = x^2 - 3$$
$$x^2 - 4x - 3 = 0$$
$$x = \frac{4 \pm \sqrt{(-4)^2 - 4(1)(-3)}}{2(1)}$$
$$= \frac{4 \pm 2\sqrt{7}}{2}$$
$$= 2 \pm \sqrt{7}$$
$$= 2 + \sqrt{7} \approx 4.646$$

Reject $2 - \sqrt{7} \approx -0.646$ since $x > 0$. Therefore the length of one side is ≈ 4.646.

47.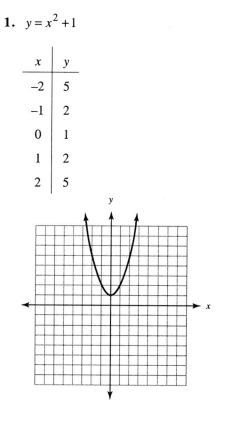

Let x be width of frame.
$$\text{Area of picture} = 140$$
$$(15 - 2x)(12 - 2x) = 140$$
$$180 - 54x + 4x^2 = 140$$
$$4x^2 - 54x + 40 = 0$$
$$2x^2 - 27x + 20 = 0$$
$$x = \frac{27 \pm \sqrt{(-27)^2 - 4(2)(20)}}{2(2)}$$
$$= \frac{27 \pm \sqrt{569}}{4}$$
$$\approx 0.787, \ 12.713$$

The width of the frame is ≈ 0.787 inches as a width of 12.713 is not possible.

49.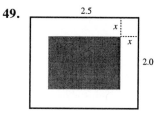

Let x be width of the frame.
$$\text{Area exposed} = 2.5$$
$$(2.5 - 2x)(2 - 2x) = 2.5$$
$$5 - 9x + 4x^2 = 2.5$$
$$4x^2 - 9x + 2.5 = 0$$
$$x = \frac{9 \pm \sqrt{(-9)^2 - 4(4)(2.5)}}{2(4)}$$
$$= \frac{9 \pm \sqrt{41}}{8}$$
$$\approx 0.325, \ 1.925$$

The width of the frame is ≈ 0.325 m or 32.5 cm.

51. Writing exercise

53. Writing exercise

Exercises 10.4

1. $y = x^2 + 1$

x	y
-2	5
-1	2
0	1
1	2
2	5

3. $y = x^2 - 4$

x	y
-2	0
-1	-3
0	-4
1	-3
2	0

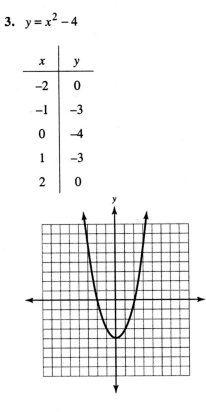

5. $y = x^2 - 4x$

x	y
0	0
1	-3
2	-4
3	-3
4	0

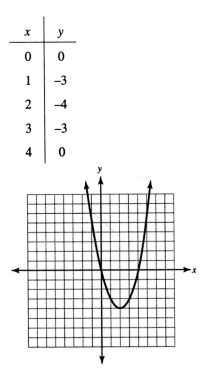

7. $y = x^2 + x$

x	y
-2	2
-1	0
0	0
1	2
2	6

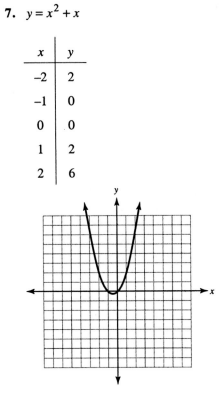

9. $y = x^2 - 2x - 3$

x	y
-1	0
0	-3
1	-4
2	-3
3	0

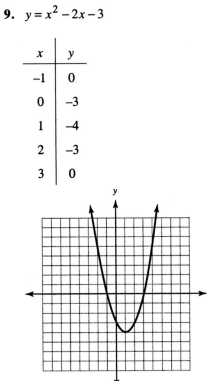

11. $y = x^2 - x - 6$

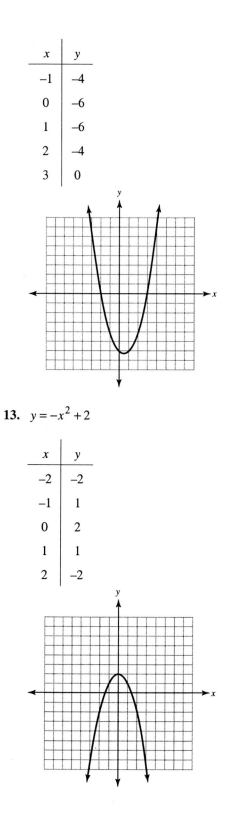

x	y
–1	–4
0	–6
1	–6
2	–4
3	0

13. $y = -x^2 + 2$

x	y
–2	–2
–1	1
0	2
1	1
2	–2

15. $y = -x^2 - 4x$

x	y
–4	0
–3	3
–2	4
–1	3
0	0

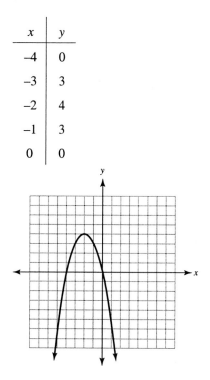

17. Parabola that opens upward, positive y-intercept, symmetric about y-axis.
(f) $y = x^2 + 1$

19. Parabola that opens downward, positive y-intercept, symmetric about y-axis.
(a) $y = -x^2 + 1$

21. Straight line with slope 2 and y-intercept 0.
(b) $y = 2x$

23. Parabola that opens downward, passes through (0, 0) and positive x-axis.
(e) $y = -x^2 + 3x$

Summary Exercises for Chapter 10

1. $x^2 = 10$
   $x = \pm\sqrt{10}$

3. $x^2 - 20 = 0$
   $x^2 = 20$
   $x = \pm 2\sqrt{5}$

5. $(x-1)^2 = 5$

$\qquad x - 1 = \pm\sqrt{5}$

$\qquad\qquad x = 1 \pm \sqrt{5}$

7. $(x+3)^2 = 5$

$\qquad x + 3 = \pm\sqrt{5}$

$\qquad\qquad x = -3 \pm \sqrt{5}$

9. $4x^2 = 27$

$\qquad x^2 = \dfrac{27}{4}$

$\qquad x = \dfrac{\pm 3\sqrt{3}}{2}$

11. $25x^2 = 7$

$\qquad x^2 = \dfrac{7}{25}$

$\qquad x = \dfrac{\pm\sqrt{7}}{5}$

13. $x^2 - 3x - 10 = 0$

$\qquad x^2 - 3x = 10$

$\qquad x^2 - 3x + \dfrac{9}{4} = 10 + \dfrac{9}{4}$

$\qquad \left(x - \dfrac{3}{2}\right)^2 = \dfrac{49}{4}$

$\qquad x - \dfrac{3}{2} = \dfrac{\pm 7}{2}$

$\qquad\qquad x = \dfrac{3}{2} \pm \dfrac{7}{2}$

$\qquad\qquad x = -2,\ 5$

15. $x^2 - 5x + 2 = 0$

$\qquad x^2 - 5x = -2$

$\qquad x^2 - 5x + \dfrac{25}{4} = -2 + \dfrac{25}{4}$

$\qquad \left(x - \dfrac{5}{2}\right)^2 = \dfrac{17}{4}$

$\qquad x - \dfrac{5}{2} = \dfrac{\pm\sqrt{17}}{2}$

$\qquad\qquad x = \dfrac{5 \pm \sqrt{17}}{2}$

17. $x^2 - 4x - 4 = 0$

$\qquad x^2 - 4x = 4$

$\qquad x^2 - 4x + 4 = 4 + 4$

$\qquad (x-2)^2 = 8$

$\qquad x - 2 = \pm 2\sqrt{2}$

$\qquad\qquad x = 2 \pm 2\sqrt{2}$

19. $x^2 - 4x = -2$

$\qquad x^2 - 4x + 4 = -2 + 4$

$\qquad (x-2)^2 = 2$

$\qquad x - 2 = \pm\sqrt{2}$

$\qquad\qquad x = 2 \pm \sqrt{2}$

21. $x^2 - x = 7$

$\qquad x^2 - x + \dfrac{1}{4} = 7 + \dfrac{1}{4}$

$\qquad \left(x - \dfrac{1}{2}\right)^2 = \dfrac{29}{4}$

$\qquad x - \dfrac{1}{2} = \dfrac{\pm\sqrt{29}}{2}$

$\qquad\qquad x = \dfrac{1 \pm \sqrt{29}}{2}$

23. $2x^2 - 4x - 7 = 0$

$\qquad 2x^2 - 4x = 7$

$\qquad x^2 - 2x = \dfrac{7}{2}$

$\qquad x^2 - 2x + 1 = \dfrac{7}{2} + 1$

$\qquad (x-1)^2 = \dfrac{9}{2}$

$\qquad x - 1 = \pm\sqrt{\dfrac{9}{2}}$

$\qquad x - 1 = \dfrac{\pm 3\sqrt{2}}{2}$

$\qquad\qquad x = 1 \pm \dfrac{3\sqrt{2}}{2}$

$\qquad\qquad x = \dfrac{2 \pm 3\sqrt{2}}{2}$

25. $x^2 - 5x - 14 = 0$

$\qquad x = \dfrac{5 \pm \sqrt{(-5)^2 - 4(1)(-14)}}{2(1)}$

$\qquad x = \dfrac{5 \pm 9}{2}$

$\qquad x = -2,\ 7$

27. $x^2 + 5x - 3 = 0$

$$x = \frac{-5 \pm \sqrt{5^2 - 4(1)(-3)}}{2(1)}$$

$$x = \frac{-5 \pm \sqrt{37}}{2}$$

29. $x^2 - 6x + 1 = 0$

$$x = \frac{6 \pm \sqrt{(-6)^2 - 4(1)(1)}}{2(1)}$$

$$x = \frac{6 \pm \sqrt{32}}{2}$$

$$x = \frac{6 \pm 4\sqrt{2}}{2}$$

$$x = 3 \pm 2\sqrt{2}$$

31. $3x^2 - 4x = 2$

$3x^2 - 4x - 2 = 0$

$$x = \frac{4 \pm \sqrt{(-4)^2 - 4(3)(-2)}}{2(3)}$$

$$x = \frac{4 \pm \sqrt{40}}{6}$$

$$x = \frac{4 \pm 2\sqrt{10}}{6}$$

$$x = \frac{2 \pm \sqrt{10}}{3}$$

33. $(x - 1)(x + 4) = 3$

$x^2 + 3x - 4 = 3$

$x^2 + 3x - 7 = 0$

$$x = \frac{-3 \pm \sqrt{3^2 - 4(1)(-7)}}{2(1)}$$

$$x = \frac{-3 \pm \sqrt{37}}{2}$$

35. $2x^2 - 8x = 12$

$2x^2 - 8x - 12 = 0$

$2(x^2 - 4x - 6) = 0$

$x^2 - 4x - 6 = 0$

$$x = \frac{-(-4) \pm \sqrt{(-4)^2 - 4(1)(-6)}}{2(1)}$$

$$x = \frac{4 \pm \sqrt{40}}{2}$$

$$x = \frac{4 \pm 2\sqrt{10}}{2}$$

$$x = 2 \pm \sqrt{10}$$

37. $5x^2 = 3x$

$5x^2 - 3x = 0$

$x(5x - 3) = 0$

$x = 0$ or $5x - 3 = 0$

$x = 0$ or $x = \dfrac{3}{5}$

39. $(x - 1)^2 = 10$

$x - 1 = \pm\sqrt{10}$

$x = 1 \pm \sqrt{10}$

41. $2x^2 = 5x + 4$

$2x^2 - 5x - 4 = 0$

$$x = \frac{5 \pm \sqrt{(-5)^2 - 4(2)(-4)}}{2(2)}$$

$$x = \frac{5 \pm \sqrt{57}}{4}$$

43. $2x^2 = 5x + 7$

$2x^2 - 5x - 7 = 0$

$(x + 1)(2x - 7) = 0$

$x + 1 = 0$ or $2x - 7 = 0$

$x = -1$ or $x = \dfrac{7}{2}$

45. $3x^2 + 6x - 15 = 0$

$3(x^2 + 2x - 5) = 0$

$x^2 + 2x - 5 = 0$

$$x = \frac{-2 \pm \sqrt{2^2 - 4(1)(-5)}}{2(1)}$$

$$x = \frac{-2 \pm \sqrt{24}}{2}$$

$$x = \frac{-2 \pm 2\sqrt{6}}{2}$$

$$x = -1 \pm \sqrt{6}$$

47. $x - 2 = \dfrac{2}{x}$

$x^2 - 2x = 2$

$x^2 - 2x - 2 = 0$

$$x = \frac{2 \pm \sqrt{(-2)^2 - 4(1)(-2)}}{2(1)}$$

$$x = \frac{2 \pm \sqrt{12}}{2}$$

$$x = \frac{2 \pm 2\sqrt{3}}{2}$$

$$x = 1 \pm \sqrt{3}$$

49. $y = x^2 + 3$

x	y
−2	7
−1	4
0	3
1	4
2	7

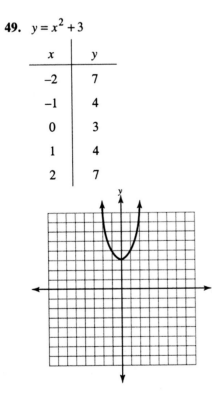

51. $y = x^2 - 3x$

x	y
−1	4
0	0
1	−2
2	−2
3	0

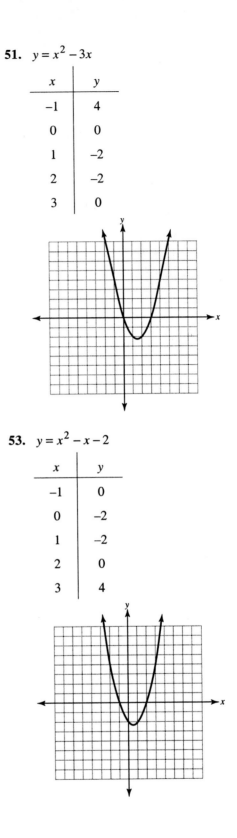

53. $y = x^2 - x - 2$

x	y
−1	0
0	−2
1	−2
2	0
3	4

55. $y = x^2 + 2x - 3$

x	y
-3	0
-2	-3
-1	-4
0	-3
1	0

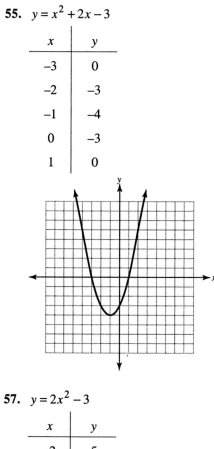

59. $y = -x^2 - 2$

x	y
-2	-6
-1	-3
0	-2
1	-3
2	-6

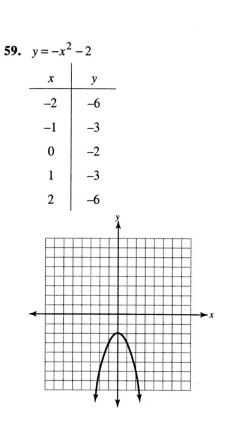

57. $y = 2x^2 - 3$

x	y
-2	5
-1	-1
0	-3
1	-1
2	5

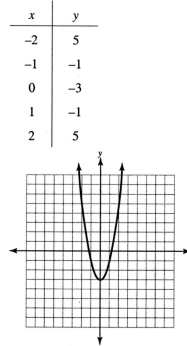